Formula One MATHS

€URO EDITION

B2

Catherine Berry ● **Margaret Bland**

Sophie Goldie ● **Julian Thomas**

Leonie Turner ● **Brandon Wilshaw**

SERIES EDITOR: **Roger Porkess**

HODDER
EDUCATION
AN HACHETTE UK COMPANY

Acknowledgements

The authors and publishers would like to thank the following companies, agencies and individuals for permission to reproduce copyright material:

Photo credits
p.3 Tony Hallas/Science Photo Library; **p.16** MEHAU KULYK/SCIENCE PHOTO LIBRARY; **p.28** ©iStockphoto.com/Thomas Pullicino; **p.29** © Roger Ressmeyer/CORBIS; **p.71** © Paul A. Souders/CORBIS; **p.106** © Ralph A. Clevenger/CORBIS; **p.107** M.C. Escher's "Symmetry drawing E112" © 2007 The M.C. Escher Company-Holland. All rights reserved. www.mcescher.com; **p130** *left* © POPPERFOTO/Alamy; *right* © POPPERFOTO; **p.148** Neil Tingle/Action Plus; **p.150** Cliff Threadgold/Life File; **p.152** © Phil Degginger/Alamy; **p.163** Wilson Bentley Digital Archives/snowflakebentley.com; **p.163** *from left to right* © doug steley/Alamy; © Nicola Vernizzi – FOTOLIA; © Shariff Che'lah – Fotolia.com; © J Marshall – Tribaleye Images/Alamy **p.176** Winning Moves UK Ltd; **p.199** Emma Lee/Life File

Every effort has been made to trace all copyright holders, but if any have inadvertently been overlooked the publishers will be pleased to make the necessary arrangements at the first opportunity.

Although every effort has been made to ensure that website addresses are correct at time of going to press, Hodder Murray cannot be held responsible for the content of any website mentioned in this book. It is sometimes possible to find a relocated web page by typing in the address of the home page for a website in the URL window of your browser.

Throughout this book, the official spelling of the words 'euro' and 'cent' have been adopted, as specified by the European Central Bank. This may be seen as departing from the usual English practice for currencies but is correct for the euro currency.

Orders: please contact Bookpoint Ltd, 130 Milton Park, Abingdon, Oxon OX14 4SB. Telephone: (44) 01235 827720. Fax: (44) 01235 400454. Lines are open 9.00–5.00, Monday to Saturday, with a 24-hour message answering service. Visit our website at www.hoddereducation.co.uk.

© 2001, 2007 Catherine Berry, Margaret Bland, Sophie Goldie, Julian Thomas, Leonie Turner, Brandon Wilshaw

First published 2001
by Hodder Education
Carmelite House,
50 Victoria Embankment,
London, EC4Y 0DZ

Euro edition published 2007

Impression number   10   9
Year   2017

Cover photo by Jacey, Debut Art
Cover design and page design by Julie Martin
Illustrations by Maggie Brand, Tom Cross, Jeff Edwards and Joe McEwan.
Typeset by Tech-Set Ltd, Gateshead, Tyne and Wear
Printed in Dubai

A catalogue record for this title is available from The British Library

ISBN: 978 0 340 94256 7

Hachette's policy is to use papers that are natural, renewable and recyclable products and made from wood grown in sustainable forests. The logging and manufacturing processes are expected to conform to the environmental regulations of the country of origin.

Introduction

This book is designed for Year 8 students and is part of a series covering Key Stage 3 Mathematics. Each textbook in the series is accompanied by an extensive Teacher's Pack including additional material. This allows the series to be used with the full ability range of students.

The series builds on the National Numeracy Strategy in primary schools and its extension into Key Stage 3. It is designed to support the style of teaching and the lesson framework to which students will be accustomed.

This book is presented as a series of double-page spreads, each of which is designed to be a teaching unit. The left-hand page covers the material to be taught and the right-hand page provides examples for the students to work through. Each chapter ends with a review exercise covering all its content. Further worksheets, tests and ICT materials are provided in the Teacher's Pack.

An important feature of the left-hand pages is the Tasks, which are printed in boxes. These are intended to be carried out by the student in mid-lesson. Their aim is twofold: in the first place they give the students practice on what they have just been taught, allowing them to consolidate their understanding. However, the tasks then extend the ideas and raise questions, setting the agenda for the later part of the lesson. Further guidance on the Tasks is available in the Teacher's Pack.

Another key feature of the left-hand pages is the Discussion Points. These are designed to help teachers engage their students in whole class discussion. Teachers should see the [?] icon as an opportunity and an invitation.

Several other symbols and instructions are used in this book. These are explained on the 'How to use this book' page for students opposite. The [symbol] symbol indicates to the teacher that there is additional ICT material directly linked to that unit of work. This is referenced in the teaching notes for that unit in the Teacher's Pack.

The order of the 25 chapters in this book ensures that the subject is developed logically, at each stage building on previous knowledge. The Teacher's Pack includes a Scheme of Work based on this order. However, teachers are of course free to vary the order to meet their own circumstances and needs.

This series stems from a partnership between Hodder Murray and Mathematics in Education and Industry (MEI). The authors would like to thank all those who helped in preparing this book, particularly those involved with the writing of materials for the accompanying Teacher's Pack.

Roger Porkess, 2007
Series Editor

How to use this book

 This symbol means that you will need to think carefully about a point. Your teacher may ask you to join in a discussion about it.

 This symbol next to a question means that you are allowed (and indeed expected) to use your calculator for this question.

 This symbol means exactly the opposite – you are not allowed to use your calculator for this question.

 This is a warning sign. It is used where a common mistake, or misunderstanding, is being described. It is also used to identify questions which are slightly more difficult or which require a little more thought. It should be read as 'caution'.

 This is the ICT symbol. It should alert your teacher to the fact that there is some additional material in the accompanying Teacher's Pack using ICT for this unit of work.

Each chapter of work in this book is divided into a series of double-page spreads – or units of work. The left-hand page is the teaching page, and the right-hand page involves an exercise and sometimes additional activities or investigations to do with that topic.

You will also come across the following features in the units of work:

Task

The tasks give you the opportunity to work alone, in pairs or in small groups on an activity in the lesson. It gives you the chance to practise what you have just been taught, and to discuss ideas and raise questions about the topic.

Do the right thing!

These boxes give you a set of step-by-step instructions on how to carry out a particular technique in maths, usually to do with shape work.

Do you remember?

These boxes give you the chance to review work that you have covered in the previous year.

Contents

Co-ordinates in all four quadrants

Look at this picture of the constellation of Cygnus.

 What does each small square represent?

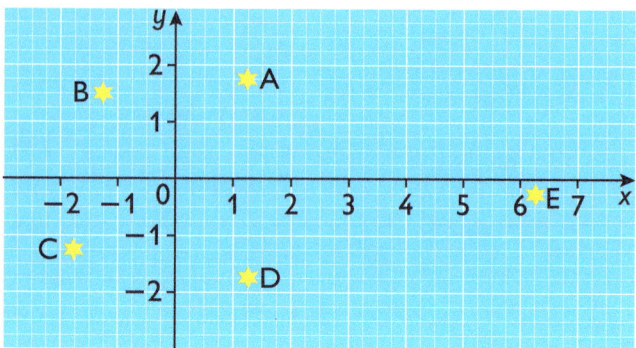

The star at E is called Vega.
The co-ordinates of Vega are $(6\frac{1}{4}, -\frac{1}{4})$.

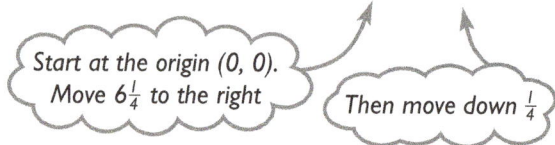

Start at the origin (0, 0). Move $6\frac{1}{4}$ to the right

Then move down $\frac{1}{4}$

 The x co-ordinate is always the first number.

 What are the co-ordinates of the other stars?

 The co-ordinates of two other stars are (2, 3) and (4, 7). A third star lies exactly half way between them. What are its co-ordinates?

 You have to be careful when working between whole numbers.
Is the x co-ordinate of A -4.5 or -3.5?
Is the x co-ordinate of B 1.1, 1.2, 1.25 or 1.4?

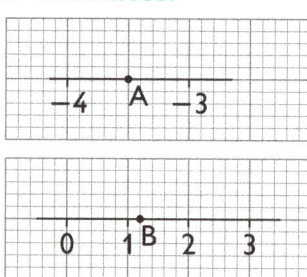

Task

1 Draw x and y axes with values from -5 to 5.

2 Plot the following points. Join them up in order.

$(-2.5, -2.5) \rightarrow (-4.5, -0.5) \rightarrow (-4.5, 1.50) \rightarrow (-3.50, 1.5)$
$\rightarrow (-3.50, 0.5) \rightarrow (-4.5, 0.5) \rightarrow (-4.5, -0.5) \rightarrow (-1.75, -0.5)$
$\rightarrow (-1.25, 0.75) \rightarrow (-0.75, 0.75) \rightarrow (-0.75, 1.75) \rightarrow (-0.25, 1.75)$
$\rightarrow (-0.25, 0.75) \rightarrow (0.5, 0.75) \rightarrow (0.5, 1.75) \rightarrow (1, 1.75)$
$\rightarrow (1, 0.75) \rightarrow (2, 0.75) \rightarrow (2.5, -0.5) \rightarrow (4.25, -0.5)$
$\rightarrow (3, -2.5) \rightarrow (-2.5, -2.5)$

What picture have you drawn?

3 Now draw a picture of your own on graph paper.
List the co-ordinates of the points in order.
Ask a friend to draw your picture.

 **What are the co-ordinates of the vertices of this shape?**

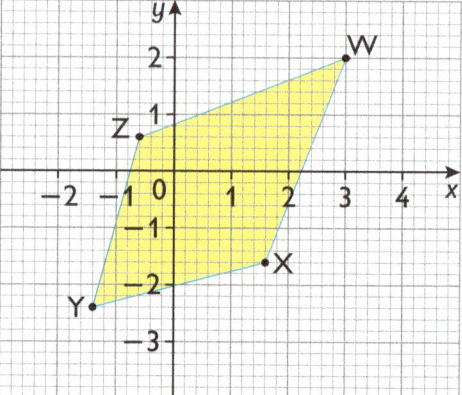

 What are the co-ordinates of the midpoint of the line joining W to the origin?

Exercise

1 These four stars form a constellation called the Square of Pegasus.

   (a) Write down their co-ordinates. Is the constellation really a square?

   (b) Draw a graph of your own with these four stars on it. Nearby, at $(-3.75, 2.5)$ is the Andromeda Galaxy. Mark it on your graph.

You can see the galaxy on a clear night. Use your graph to help you find it.

2 Draw axes with the values for both x and y from -5 to $+5$.
Plot these points and join them in order:

     $A(-3, 1)$, $B(0, 5)$ and $C(4, 2)$

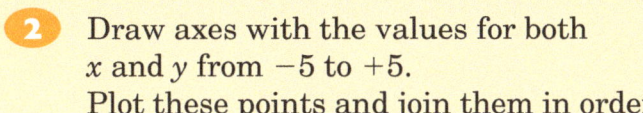

Andromeda Galaxy, M31

These are three vertices of the square ABCD.

   (a) What are the co-ordinates of D?

   (b) What are the co-ordinates of the midpoint of each side of the square?

3 Draw axes with the values for both x and y from -5 to $+5$.
Plot these points and join them in order:

     $X(-2.6, -3.8)$, $Y(-0.2, 3.2)$, $Z(4.6, 3.2)$

These are three vertices of the parallelogram WXYZ.

   (a) What are the co-ordinates of W?

   (b) What are the co-ordinates of the midpoints of **(i)** XZ **(ii)** YW?

4 A mirror is standing on the x-axis.
Two ants Bill and Ben are standing at $(0.75, -2.6)$ and $(-0.5, 3.4)$.
They look straight at the mirror.
Give the co-ordinates of their reflections.

5 Draw axes with the values for both x and y from -6 to $+6$.
Plot these points and join them in order:

     $A(-5.6, 0)$, $B(-2.2, 3.4)$, $C(1.2, 3.4)$, $D(4.6, 0)$, $E(1.2, -3.4)$

A sixth point, F, is needed to make a symmetrical hexagon.
FE is of equal length to BC and parallel to BC.

   (a) What are the co-ordinates of F?
Add F to your graph and complete the hexagon.

   (b) How many lines of reflection symmetry has the hexagon?

   (c) Are all the sides the same length? Are all the angles equal?

The equation of a line

Rachel and Steve are playing a game of *4 in a line* using co-ordinates.
They take it in turns to plot points.

The winner is the first to make a straight line of four points either horizontally, vertically or diagonally. The points must be next to each other.

Here is their game. It is Rachel's turn.

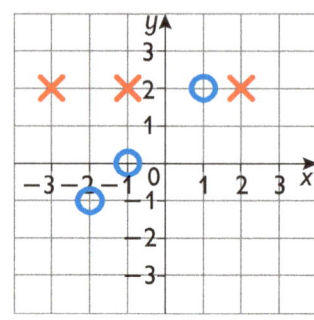

Rachel: **O**
Steve: **✕**

? **Where must Rachel go so that she gets four in a line?**

Rachel writes down the co-ordinates of her line: (–2, –1)
(–1, 0)
(0, 1)
(1, 2)

? **What pattern do you notice?**

Rachel notices that the *y* co-ordinate is always one more than the *x* co-ordinate.
She writes:

$$\text{The equation of my line is } y = x + 1$$

? **What are the co-ordinates of three other points that would also lie on the same line?**

In another game the co-ordinates of Steve's winning line are: (–1, 3)
(0, 2)
(1, 1)
(2, 0)

? **What pattern do you notice?**

Steve notices that the *x* co-ordinate and the *y* co-ordinate always add up to 2.
Steve writes:

$$\text{The equation of my line is } x + y = 2$$

Task

Play some games of *4 in a line*.
Write down the co-ordinates of the winning lines.
Find an equation for each winning line.

? **Does every winning line have an equation?**
What about horizontal and vertical lines?
Is the equation of a line true for all points on it?

Exercise

1 Write down the co-ordinates of the points marked on these lines. Find the equation for each line.

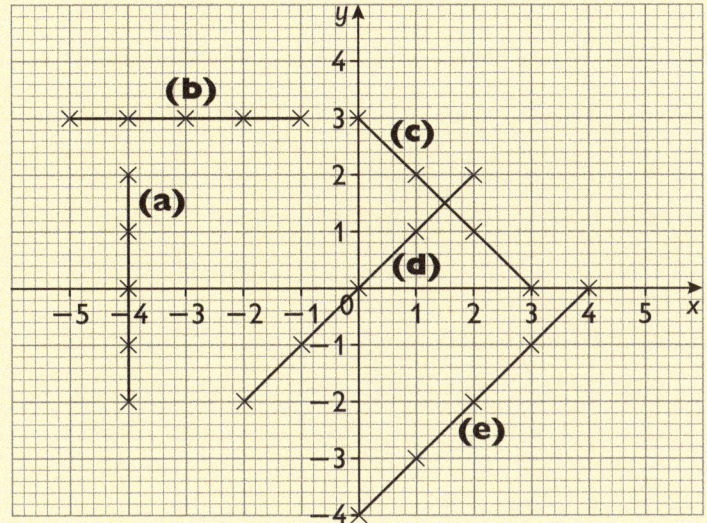

2 Complete the missing co-ordinates for the following lines. Find an equation for each line.

(a)	**(b)**	**(c)**	**(d)**
$(0, 0)$	$(6, 0)$	$(-3, -1)$	$(-1, 8)$
$(1, 1)$	$(5, 1)$	$(-2, 0)$	$(0, 7)$
$(2, 2)$	$(4, 2)$	$(-1, 1)$	$(1, 6)$
$(3, ?)$	$(3, ?)$	$(0, ?)$	$(?, 5)$
$(?, 5)$	$(?, 5)$	$(?, 6)$	$(3, ?)$
$(8.6, ?)$	$(2.4, ?)$	$(?, 2.3)$	$(?, 4.1)$

3 Write down the co-ordinates of four points which lie on each of the following lines:

(a) $x + y = 5$ **(b)** $y = x + 3$

(c) $y = x - 2$ **(d)** $x + y = 10$

Activity Use a graphic calculator to draw this picture.

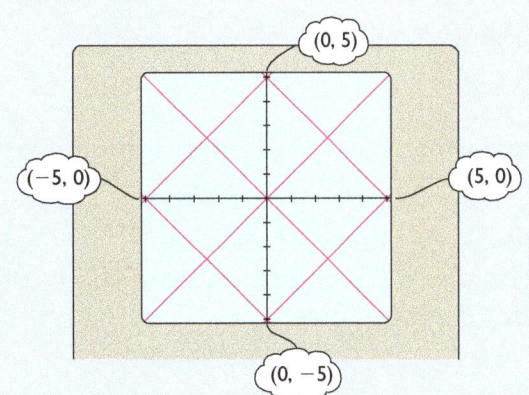

Drawing straight line graphs

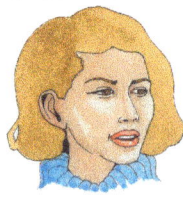

Jamie, draw the line
$y = 2x + 1$

How do I do that?

" Do the right thing!

Step 1 Choose numbers for x.

So choose $-3, -2, -1,$
$0, 1, 2, 3$ for x.

Step 2 Make a table.
Use the x co-ordinates to
work out the y co-ordinates.

$x = 3$
So $y = 2 \times 3 + 1 = 7$

x	-3	-2	-1	0	1	2	3
$2x$		-4		0			6
$+1$	$+1$	$+1$	$+1$	$+1$			$+1$
$y = 2x + 1$		-3		1			7

? **What are the rest of the y co-ordinates?**

How can you tell how big you need to draw your axes?

Step 3 Draw your axes and plot all the
points you have found.

Step 4 Join up all your points.

Notice the scales on the x axes and the y axes.

? **What does each small square represent on each axes?**

? **What is the value of y when $x = 2.2$?**
What is the value of x when $y = 4$?

The point where the line crosses the y-axes
is called the **y-intercept**.

? **What are the co-ordinates of the y-intercept for the line $y = 2x + 1$?**

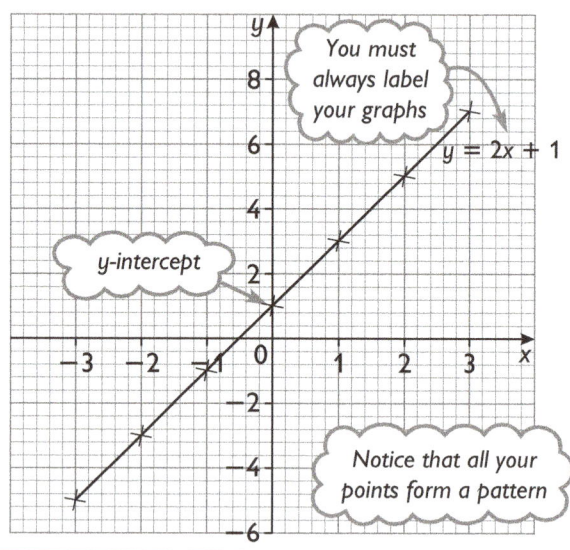

You must always label your graphs

$y = 2x + 1$

y-intercept

Notice that all your points form a pattern

Task

Draw all of these lines on the same axes.

1 $y = 2x - 1$ **2** $y = 2x$ **3** $y = 2x - 3$ **4** $y = 2x + 4$

Make sure you label each line.
Write down the co-ordinates of the y-intercept for each line.
What do you notice?

? **What are the co-ordinates of the y-intercept of the line $y = 2x + 5$?**
How can you tell?

Exercise

1 Use the following steps to work out each part of this question.

  (i) Copy and complete the table.
  (ii) Write down the largest and smallest values of x and y.
  Draw a pair of axes choosing a suitable scale.
  (iii) Draw the graph of each line.
  (iv) Write down the co-ordinates of the y-intercept of each line.

(a)

x	−3	−2	−1	0	1	2	3
+2	+2	+2	+2	+2	+2	+2	+2
$y = x + 2$	−1			2			5

(b)

x	−3	−2	−1	0	1	2	3
$3x$	−9			0		6	
−1	−1			−1		−1	
$y = 3x - 1$	−10			−1		5	

(c)

x	−3	−2	−1	0	1	2	3
$4x$	−12		−4			8	
−2	−2		−2			−2	
$y = 4x - 2$	−14		−6			6	

2 Find the co-ordinates of five points which lie on:

  (i) $y = x + 2$ **(ii)** $y = 2 - x$

Use your points to draw the graphs of each line on the same axes.
Label each line.

  (a) Write down the co-ordinates of the y-intercept for each line.

  (b) What is the angle between the two lines?

  (c) Write down the co-ordinates of the point where the two lines cross.

  (d) Are there any other points which lie on both lines?

Investigation

Which of the points marked lie on the following lines:

$x + y = 0$ $y = x - 1$
$y = x$ $y = 3$
$x = 3$ $x = 0$
$y = -2$ $y = 0$

Which points lie on more than one of these lines?

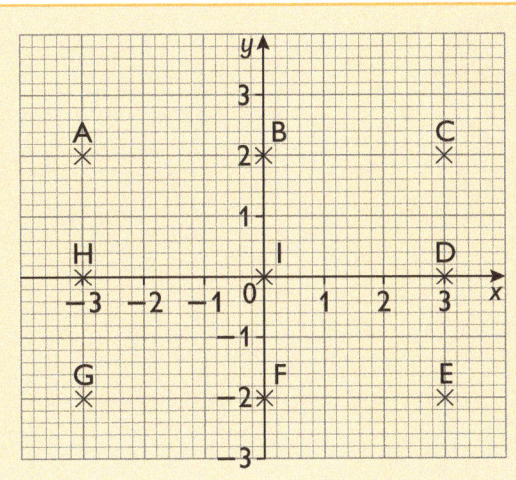

The gradient of a line

 Look at these two graphs.
Which line is steeper?

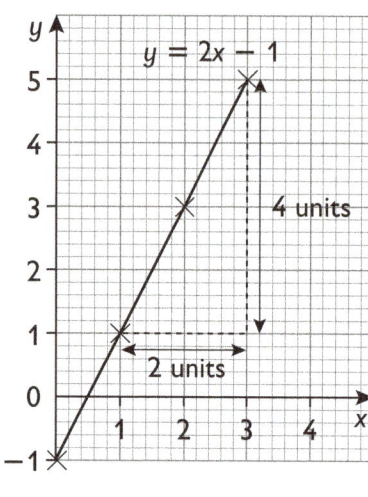

Figure 1

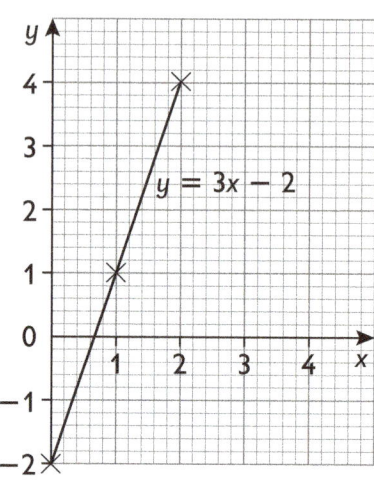

Figure 2

In mathematics we use the word **gradient** to mean the steepness of a line.
Look at Figure 1.

To find the gradient draw a triangle underneath the line.

Change in x is 2

The points make a right-angled triangle

Then work out the change in x and the change in y.

The gradient of the line is:

Change in y is 4

$$\text{Gradient} = \frac{\text{change in y}}{\text{change in x}} = \frac{4}{2} = 2$$

 What is the gradient of the line $y = 3x - 2$?
Which is the steeper line?

 Task

Find the co-ordinates of five points which lie on each of the following lines:

1 $y = x$ **2** $y = 2x$ **3** $y = 3x$

Use your co-ordinates to draw the graphs of each line on the same pair of axes.
Calculate the gradient in each case.

 What do you notice about the gradients of the lines?
What is the gradient of the line $y = 20x$?

 Be careful to check x and y scales.

Exercise

1 Find the gradients of the following lines:

(a)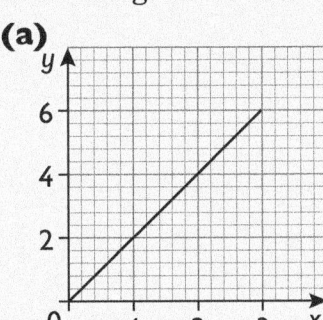

(b)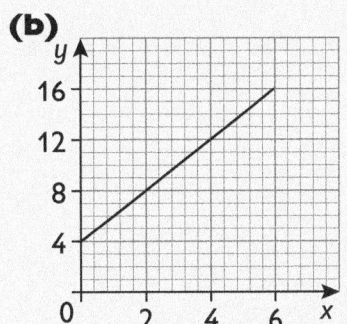

(c)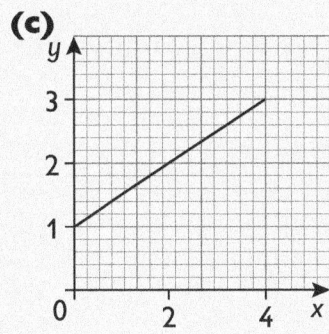

2 Draw the graphs of the following lines on the same axes.
Use the same scales for the x and y axes.
Take values of x from -4 to $+4$.

(i) $y = x$　　　　(ii) $y = 3 - x$　　　　(iii) $y = x + 3$

Label each line.

(a) Which lines are parallel?

(b) What is the gradient of the two parallel lines?

(c) Which lines have the same y-intercept?

(d) What are the co-ordinates of the points where one line crosses the other two?

(e) Which lines are perpendicular?

3 (a) Which line is steeper $y = 3x - 12$ or $y = 2x + 20$?

(b) What is the gradient of the line $y = 10x$?

4 Here are the graphs of the lines

(i) $y = x - 2$　and　(ii) $y = 2x$.

(a) Find the gradient of each line.

(b) Which is the steeper line?

(c) Find the y-intercept of each line.

(d) Write down the co-ordinates of the point where the two lines cross.

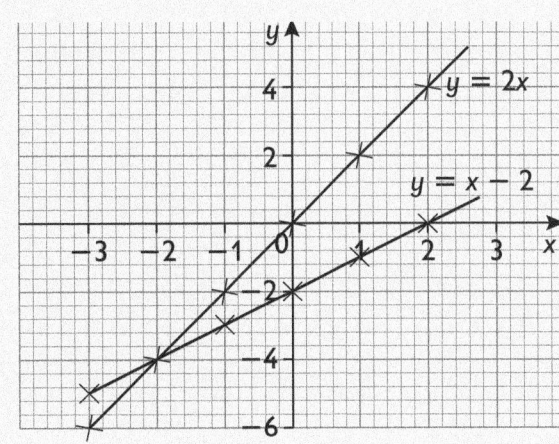

Finishing off

Now that you have finished this chapter you should be able to:

- plot co-ordinates in all four quadrants
- find the equation of a line
- draw a line given its equation
- understand what the y-intercept is
- understand what is meant by the gradient of a line
- find the gradient of a line.

Review exercise

1 Write down the co-ordinates of the points A, B, C and D.

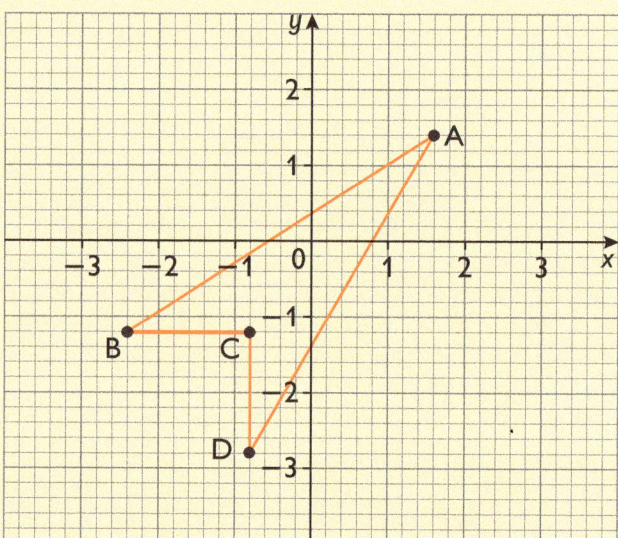

2 For each line a, b, c and d:

(a) Write down the co-ordinates of each point marked.

(b) Find the equation of the line.

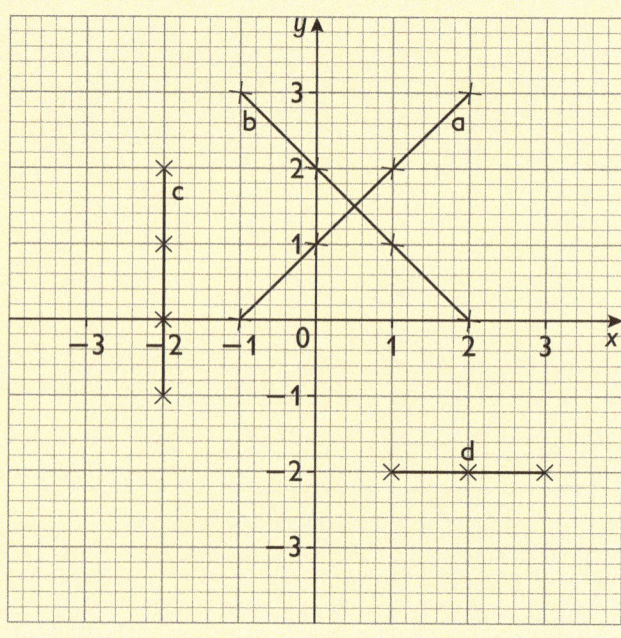

3 **(a)** Copy and complete this table for $y = 3x - 2$.

x	−3	−2	−1	0	1	2	3
$3x$		−6		0			9
−2		−2		−2			−2
$y = 3x - 2$		−8		−2			7

(b) Make out a table for $y = 4x + 1$.
Take x from −3 to +3.
Complete the table.

(c) **(i)** Draw the graphs of $y = 3x - 2$ and $y = 4x + 1$ on the same pair of axes.
Label each line clearly.

(ii) Write down the co-ordinates of the y-intercepts for both lines.

(iii) What are the co-ordinates of the point which lies on both lines?

(iv) Find the gradient of each line.

4 Find the gradient and y-intercept of each of these lines:

(a)
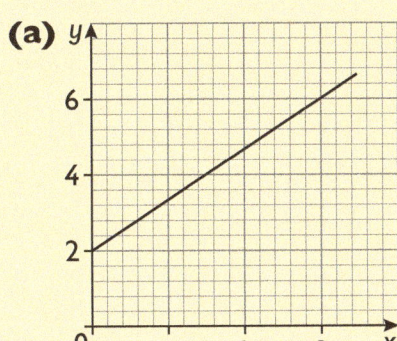

(b)
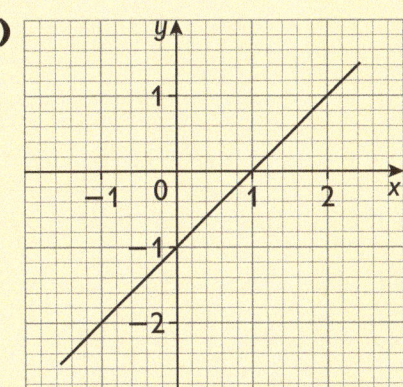

5 Which is the steeper $y = 5x - 3$ or $y = 4x + 20$?
Give a reason for your answer.

Investigation Draw the following graphs on the same axes:

(i) $y = 2x + 2$ **(ii)** $y = x + 4$ **(iii)** $y = 2x - 3$

(iv) $y = 4 + 2x$ **(v)** $y = x + 2$ **(vi)** $y = x - 3$

(a) Which lines have the same gradient?
What do you notice about their equations?

(b) Which pairs of lines have the same y-intercept.
What do you notice?

(c) A line has gradient 3 and y-intercept −5.
What is its equation?

2 Numbers

Decimals revisited

Look at these weights. In grams the numbers are large. It is easier to think of the weights in kilograms.
There are 1000 grams in a kilogram.

Prize-winning vegetables!

LEEK 3500 grams from Northumberland

POTATO 2930 grams from Sedley

 Work out

(a) $3500\,\text{g} \times 0.001$

(b) $2930\,\text{g} \times 0.001$

(c) $3500\,\text{g} \times \frac{1}{1000}$

(d) $2930\,\text{g} \times \frac{1}{1000}$

To change grams to kilograms divide by 1000.

What do you notice?
What is 0.001 as a fraction?

 Task

Work out, giving your answers as decimals

1 (a) (i) 6×0.1 **(ii)** 8×0.1 **(iii)** 3×0.1 **(iv)** 27×0.1

(b) (i) $6 \times \frac{1}{10}$ **(ii)** $8 \times \frac{1}{10}$ **(iii)** $3 \times \frac{1}{10}$ **(iv)** $\frac{1}{10} \times 27$

(c) (i) $\frac{6}{10}$ **(ii)** $\frac{8}{10}$ **(iii)** $\frac{3}{10}$ **(iv)** $\frac{27}{10}$

2 (a) (i) 7×0.01 **(ii)** 3×0.01 **(iii)** 36×0.01

(b) (i) $7 \times \frac{1}{100}$ **(ii)** $3 \times \frac{1}{100}$ **(iii)** $36 \times \frac{1}{100}$

(c) (i) $\frac{7}{100}$ **(ii)** $\frac{3}{100}$ **(iii)** $\frac{36}{100}$

3 (a) (i) 6×0.001 **(ii)** 56×0.001 **(iii)** 9×0.001

(b) (i) $6 \times \frac{1}{1000}$ **(ii)** $56 \times \frac{1}{1000}$ **(iii)** $9 \times \frac{1}{1000}$

(c) (i) $\frac{6}{1000}$ **(ii)** $\frac{56}{1000}$ **(iii)** $\frac{9}{1000}$

What do you notice about your results?

Not all changes of units involve the decimal system.
Time is not decimal.

Examples 0.1 of an hour = 0.1 of 60 minutes   78 minutes = $\frac{78}{60}$ hours
= 0.1 × 60 mins   = 1.3 hours
= 6 mins   or 1 hour and 18 minutes

 What other changes of units do not involve the decimal system?

Exercise

1 Write down
- **(a)** 7×0.1
- **(b)** 7×0.01
- **(c)** 7×0.001
- **(d)** 17×0.1
- **(e)** 17×0.01
- **(f)** 17×0.001
- **(g)** 415×0.1
- **(h)** 415×0.01
- **(i)** 415×0.001

2 Write down
- **(a)** 3×0.1
- **(b)** 3×0.01
- **(c)** 3×0.001
- **(d)** 30×0.1
- **(e)** 30×0.01
- **(f)** 30×0.001
- **(g)** 300×0.1
- **(h)** 3000×0.1
- **(i)** $30\,000 \times 0.01$

3 The weight of the world's largest lemon was 2130 grams.
Change this to kilograms.

4 The weight of the world's largest watermelon was 52 980 grams.
Change this to kilograms.

5 The Malaysian orchid can grow to 760 cm tall.
Change this height to metres.

6 The world's largest cactus found in Arizona was
1767 cm tall.
Change this height to metres.

World's largest cactus

7 Change these times to hours and minutes.
- **(a)** 6.1 hours
- **(b)** 8.5 hours
- **(c)** 3.4 hours
- **(d)** 5.2 hours
- **(e)** 7.3 hours
- **(f)** 1.8 hours

8 Change these times to minutes and seconds.
- **(a)** 1.5 mins
- **(b)** 3.1 mins
- **(c)** 6.2 mins
- **(d)** 2.8 mins
- **(e)** 9.7 mins
- **(f)** 4.3 mins

Activity

1 Work out the following questions. You may use your calculator.
- **(a)** $\dfrac{7}{0.1}$
- **(b)** $\dfrac{7}{0.01}$
- **(c)** $\dfrac{7}{0.001}$
- **(d)** $\dfrac{5}{0.1}$
- **(e)** $\dfrac{5}{0.01}$
- **(f)** $\dfrac{5}{0.001}$
- **(g)** $\dfrac{50}{0.1}$
- **(h)** $\dfrac{50}{0.01}$
- **(i)** $\dfrac{50}{0.001}$
- **(j)** $\dfrac{3}{0.1}$
- **(k)** $\dfrac{30}{0.1}$
- **(l)** $\dfrac{30}{0.01}$

Look carefully at your answers to find a non-calculator method.

2 Work out

- **(a)** $\dfrac{8}{0.1}$
- **(b)** $\dfrac{8}{0.01}$
- **(c)** $\dfrac{8}{0.001}$
- **(d)** $\dfrac{6}{0.1}$
- **(e)** $\dfrac{60}{0.1}$
- **(f)** $\dfrac{60}{0.01}$

Using scales

Sarah is planning a new kitchen. It is rectangular and has one door 0.7 m wide.

These are her measurements for the length

2 m   3 m   4 m   5 m

and the width.

3 m   4 m   5 m   6 m

1 Using squared paper, make a scale drawing of the kitchen. The kitchen must contain

Use a scale of 2 cm to represent 1 m.

A work surface 1.7 m long and 0.7 m wide.

A table 1 m long and 0.8 m wide.

A cooker 1.2 m long and 0.7 m wide.

A washing machine, a fridge and a freezer all 0.6 m long and 0.7 m deep.

2 Cut out rectangles, drawn to scale, to represent all of the items above.

3 Now design Sarah's kitchen.

You worked with scales in the Task.
You must always be careful. Check what each division means below.

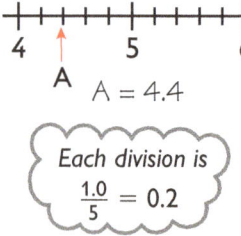

A = 4.4

Each division is
$\frac{1.0}{5} = 0.2$

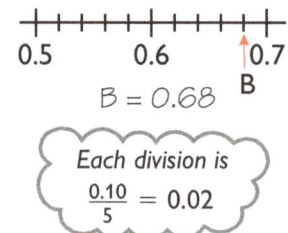

B = 0.68

Each division is
$\frac{0.10}{5} = 0.02$

This picture shows the dials in Sarah's car.

What are the readings?

What do all the dials measure?

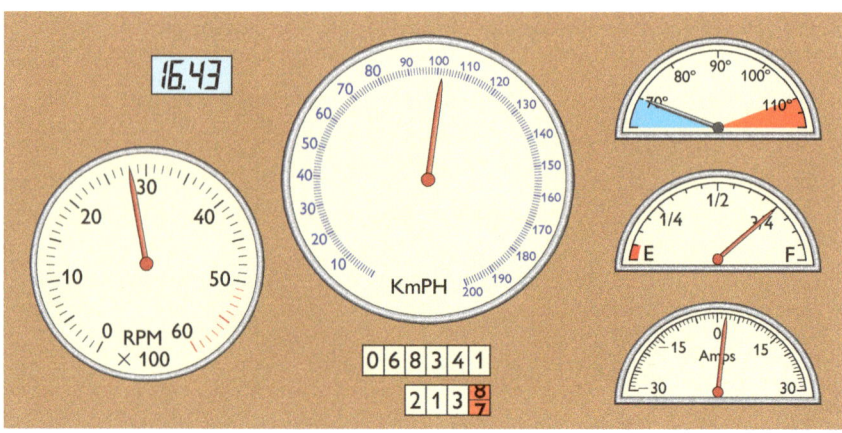

Exercise

1 The cockpit of an aeroplane is full of dials.

(a) An *altimeter* shows the height of the aeroplane in 1000 m.

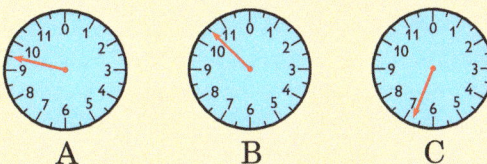

A   B   C

At what heights are these aeroplanes flying?

(b) A *speedometer* shows the airspeed in kilometres per hour.

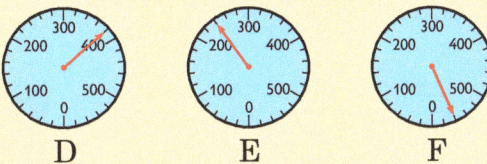

D   E   F

At what speeds are these aeroplanes flying?

(c) Look at the *fuel gauge*.
A full tank contains 3000 kg.
How much fuel is there when the pointer is at

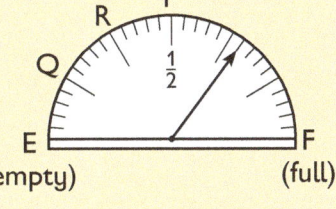

   (i) P   **(ii)** Q   **(iii)** R?

   (iv) How much fuel is the pointer in the
          diagram showing?

2 What numbers are shown on the number line below?

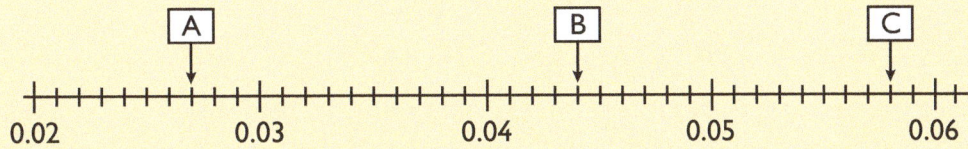

3 What numbers are shown on the number line below?

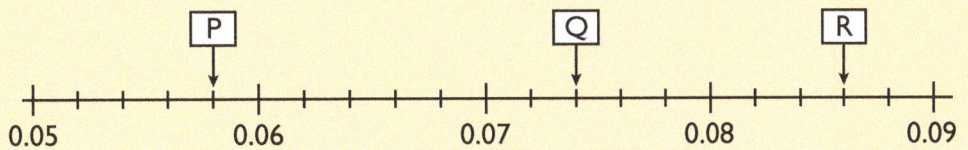

4 Copy the number line below and mark on these numbers.

   (a) 0.18   **(b)** 0.24   **(c)** 0.27   **(d)** 0.15

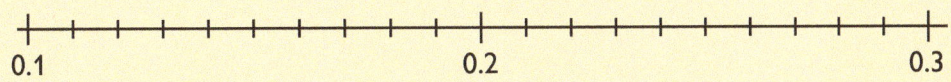

5 Copy the number line below and mark on these numbers.

   (a) 0.18   **(b)** 0.24   **(c)** 0.32   **(d)** 0.38

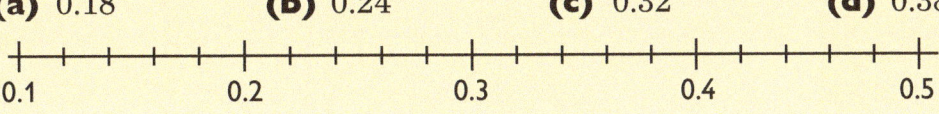

Ordering

? **Look at these pairs of numbers. In each case say which is the larger and which the smaller.**

(a) 0.0909 and 0.1001

(b) 10^2 and 10^3

(c) One million and 100×1000

(d) 2^{10} and 10^3

(e) $\dfrac{1}{10^9}$ and $\dfrac{1}{10^8}$

(f) $\dfrac{5}{1000}$ and $\dfrac{1}{500}$

Task

The planets in the solar system go round the Sun. Their orbits are nearly circles (except for Pluto).

Here are their distances from the Sun, in millions of kilometres.

Jupiter	780	Mercury	58
Earth	150	Neptune	4500
Uranus	2870	Venus	108
Mars	230	Saturn	1430

1 Put the planets in order, starting with the one nearest the Sun.

2 Which planets are between Earth and Saturn?

3 Which planet is about twice as far from the Sun as Saturn?

4 How many planets are less than one thousand million kilometres from the Sun?

5 There is a large gap between Mars and Jupiter.
How many million kilometres is the gap?
(There are a lot of small planets, called asteroids, in the gap.)

? **Which planet is about 3000 million km from the Sun?**

The number 3000 has been rounded to the nearest 1000.

? **Round the distances of Mars and Venus from the Sun to the nearest 100 million km.**

? **A friend asks you for the distances of the planets from the Sun in 'round numbers'.
How would you answer your friend for each planet?
Which would you give to the nearest 10 million, the nearest 100 million, the nearest 1000 million km?
What about Earth and Neptune?**

Exercise

1 The list below shows temperatures (in °C) of 8 places at noon one day in January. Arrange them in order starting with the coldest.

(a) Felixstowe   3.9   **(b)** Penzance   4.2   **(c)** Keswick 3.7

(d) Margate   4.6   **(e)** Greenock   3.8   **(f)** Durham   0

(g) Aberdeen   2.4   **(h)** Birmingham   3.5

2 Some comets pass the Earth only once, some return at regular intervals. Arrange the orbital periods of these comets in order, smallest first.

Borelly	6.8 years	D'Arrest	6.2 years
Encke	3.3 years	Faye	7.4 years
Finlay	6.9 years	Pons-Winnecke	6.3 years

3 Here are the dates of the reigns of some English kings.

Work out for how long each king reigned.
Place them in order, shortest time first.

William I	1066–1087	William II	1087–1100
Henry I	1100–1135	Stephen	1135–1154
Henry II	1154–1189	Richard I	1189–1199
John	1199–1216	Henry III	1216–1272

4 **(a)** Write the following as decimals

(i) 0.6×100 **(ii)** $\frac{12}{25}$ **(iii)** $\frac{400}{500}$ **(iv)** 0.071×100

(b) Arrange your answers in order of size, smallest first.

5 **(a)** Work out the following
(i) 10^3 **(ii)** 10^5 **(iii)** 10^6 **(iv)** 10^2 **(v)** 10^7 **(vi)** 10^4 **(vii)** 10^1

(b) Arrange your answers giving the smallest first.

(c) Write out the list again, in the same order as part (b) but using the index notion of part (a).

(d) What do you notice about the order of your answers?

6 **(a)** Write each of these distances in metres.

(i) 50 km   **(ii)** $6\,000 \times \frac{1}{10}\,\text{m}$ **(iii)** 700 000 cm
(iv) 80 000 000 mm   **(v)** $5\frac{1}{2}\,\text{km}$

(b) Write the distances in order, smallest first.

7 The speed of sound is 340 metres per second. This is known as *Mach One*.
(a) Write these speeds in metres per second.
(i) Storm Missile   Mach 0.7
(ii) Tornado Fighter   45 000 centimetres per second
(iii) Jetstream aeroplane   7 kilometres per minute
(iv) Nimrod   Mach 0.3

(b) Place them in order of size, fastest first.

Range of values

Lindsay is planting forget-me-nots in her garden.
Written on the packet is

Seeds should be planted at a depth of 1–2 cm but not deeper than 2 cm

This can be represented on a number line by solid circles.

It is written as 1 cm ≤ depth ≤ 2 cm

This *includes* 1 cm and 2 cm in the range of values.

You can write this as l ≤ d ≤ 2.

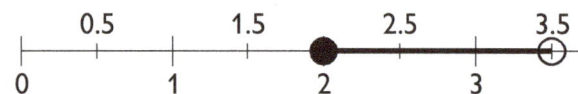

Thin out when at least 2 cm tall, but before they are 3.5 cm

This is written as 2 cm ≤ height < 3.5 cm
The 2 cm is included but the 3.5 cm is not.

This is shown on the number line below.

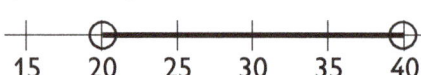

In good conditions the plants should grow to over 20 cm but less than 40 cm

This is written as 20 cm < height < 40 cm.
Neither the 20 cm nor the 40 cm is included in the range of values. This is shown on the number line by leaving circles hollow

 Task

Some flowers are listed in question 3. Sort them into classes according to their heights.
Copy the tally chart and complete it up to 1.65 m.

Height in metres	Tally	Frequency
$0 < h \leq 0.15$		
$0.15 < h \leq 0.30$		
$0.30 < h \leq 0.45$		
$0.45 < h \leq 0.60$		

? **Which class of heights occurs most frequently?**

These equalities hold for a number N.

$15 \leq N < 20$ $18 < N < 22$
$14 < N \leq 19$ $16 \leq N \leq 20$

 What is N? Do you need all these inequalities to find N?

Exercise

1 Show the following ranges of values on number lines.

(a) $4 < x < 6$ (b) $5 \leqslant x < 8$

(c) $6 \leqslant x \leqslant 10$ (d) $-1 < x \leqslant 2$

(e) $2.1 < x < 2.4$ (f) $6.2 \leqslant x < 6.8$

(g) $5.3 \leqslant x \leqslant 5.7$ (h) $4.5 < x \leqslant 5.5$

(i) $0.15 < x \leqslant 0.2$ (j) $0.24 \leqslant x < 0.3$

(k) $0.03 < x < 0.05$ (l) $0.14 \leqslant x \leqslant 0.23$

2 Write down two numbers in each of the following ranges.

(a) $2 \leqslant x \leqslant 4$ (b) $2 < x < 3$

(c) $2.3 < x < 2.7$ (d) $2.13 < x < 2.14$

(e) $5.13 \leqslant x \leqslant 5.18$ (f) $5.13 \leqslant x \leqslant 5.14$

(g) $0.07 < x \leqslant 0.09$ (h) $0.07 < x < 0.08$

(i) $4.327 \leqslant x < 4.413$ (j) $4.327 \leqslant x \leqslant 4.328$

3 Packets of seeds state the average height of each type of flower. The heights are given below in metres.

Alyssum	0.25	Antirrhinum	0.17
Aquilegia	0.80	Aster (dwarf)	0.15
Begonia	0.15	Californian Poppy	0.30
Carnation	0.55	Dahlia	1.15
Delphinium	1.20	Diascia	0.25
Gaillardia	0.35	Geranium	0.30
Godetia	0.45	Gypsophila	0.50
Hollyhock	1.65	Lace Flowers	0.65
Larkspur	0.60	Lobelia	0.15
Marigold	0.35	Mimulus	0.15
Nasturtium (dwarf)	0.25	Pansies	0.20
Petunia	0.15	Phlox	0.30
Primrose	0.15	Poppy	0.90
Rock Rose	0.30	Rudbeckia	0.45
Statice	0.45	Stocks	0.70
Virginian Stocks	0.15	Wallflower	0.35

(a) Which of these flowers fall into the range $0.4 < h \leqslant 0.6$?

Statice   Carnation   Lace Flowers

(b) Which of these flowers fall into the range $0.2 < h \leqslant 0.3$?

Alyssum   Pansies   Rock Rose

(c) Which of these flowers fall into the range $0.35 < h \leqslant 0.55$?

Wallflower   Gypsophila   Carnation

Finishing off

Review exercise

1 The heights of five Russian dolls are

9.6 cm   3.6 cm   7.4 cm   13.5 cm   5.2 cm

Arrange them in descending order. Start with the largest.

2 The times of the first five athletes in a women's 100 metres race were:

11.08 s   10.79 s   10.88 s   10.77 s   10.81 s

Arrange these in ascending order. Start with the smallest.

3 The times of the first five athletes in a men's 100 metres race were:

10.03 s   9.95 s   10.4 s   10.2 s   9.93 s

Arrange these in ascending order. Start with the smallest.

4 The diameters of the nine planets in the Solar System are

Mercury	4.9×10^3 km	Venus	1.21×10^4 km
Earth	1.2756×10^4 km	Mars	6.8×10^3 km
Jupiter	1.43×10^5 km	Saturn	1.2×10^5 km
Uranus	5.2×10^4 km	Neptune	4.8×10^4 km
Pluto	3.0×10^3 km		

Arrange these diameters in descending order. Start with the largest.

5 What are the values of A, B and C on this number line?

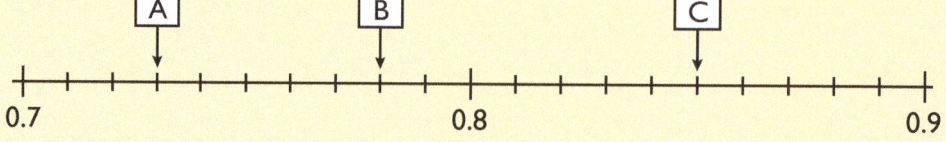

6 What are the values of X, Y and Z on this number line?

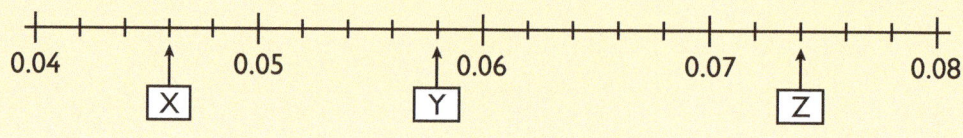

7 The trees in a plantation are all over 25 m tall and under 35 m.
Write an inequality to describe this and show it on a number line.

8 Stocks grow to a height of over 60 cm and less than 75 cm.
Write this as an inequality and draw a number line to show it.

9 Work out as decimals

(a) $\frac{17}{100}$ (b) $\frac{1.7}{100}$ (c) $\frac{0.17}{100}$

(d) 17×0.01 (e) 1.7×0.01 (f) 0.17×0.01

(g) $17 \times \frac{1}{100}$ (h) $1.7 \times \frac{1}{100}$ (i) $0.17 \times \frac{1}{100}$

10 Work out as decimals

(a) $\frac{236}{1000}$ (b) $\frac{9}{1000}$ (c) $\frac{16}{1000}$

(d) 236×0.001 (e) 9×0.001 (f) 16×0.001

(g) $236 \times \frac{1}{1000}$ (h) $9 \times \frac{1}{1000}$ (i) $16 \times \frac{1}{1000}$

11 What happens when you multiply a number by another number which is less than 1?

12 Write down

(a) 6×10 (b) 0.6×100 (c) 0.06×1000

(d) $\frac{6}{0.1}$ (e) $\frac{0.6}{0.1}$ (f) $\frac{0.06}{0.001}$

13 Write down

(a) 2×1000 (b) 4×1000 (c) 7.3×1000

(d) $\frac{2}{0.001}$ (e) $\frac{4}{0.001}$ (f) $\frac{7.3}{0.001}$

14 What happens when you divide a number by another number which is less than 1?

15 (a) Using the formula *time = distance ÷ speed* work out the times in hours for these journeys.
    (i) Bristol to Birmingham   144 km at 80 km/h
    (ii) Nottingham to Leicester   45 km at 50 km/h
    (iii) Cambridge to Oxford   128 km at 40 km/h

(b) Change all these times to hours and minutes.

3 Angles

Types of angle

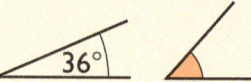

Task

Look at these diagrams then copy and complete the table below.

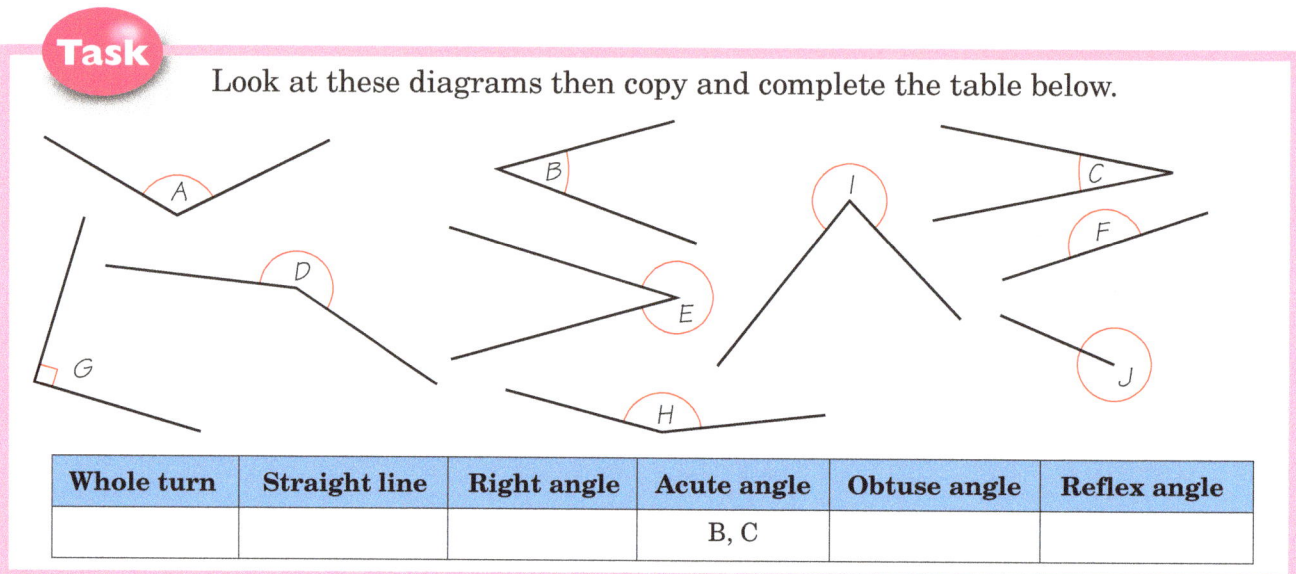

Whole turn	Straight line	Right angle	Acute angle	Obtuse angle	Reflex angle
			B, C		

? What do acute, obtuse and reflex each mean?

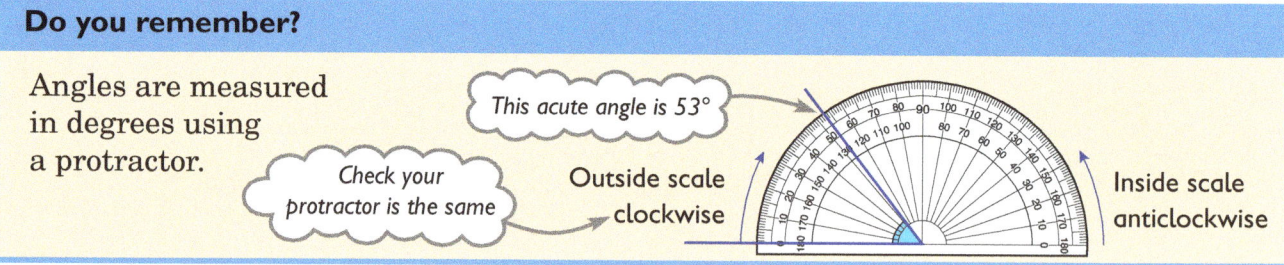

This acute angle is 53°

Check your protractor is the same

Outside scale clockwise

Inside scale anticlockwise

? What is the size of (a) the largest acute angle (b) the smallest obtuse angle?

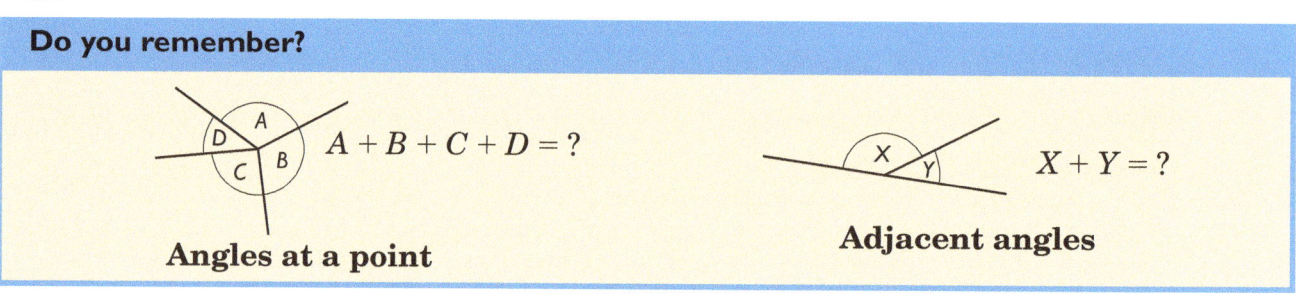

$A + B + C + D = ?$

$X + Y = ?$

Angles at a point

Adjacent angles

? How do you use a protractor to measure or draw a reflex angle?

Exercise

1. How many right angles are
   (a) on a straight line   (b) in a whole turn   (c) in 270°?

2. Which of these angles are (a) acute (b) obtuse (c) reflex?

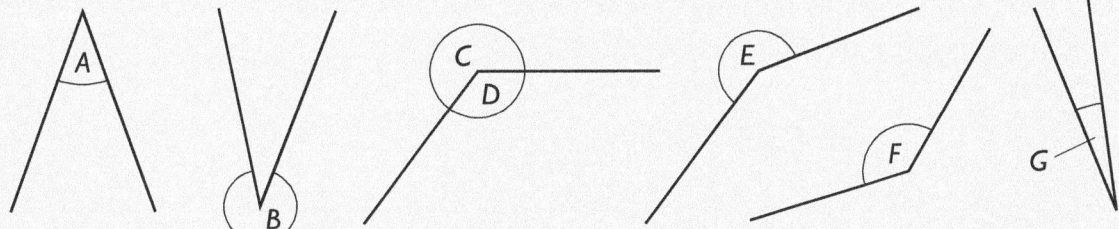

3. Measure these angles as accurately as you can.

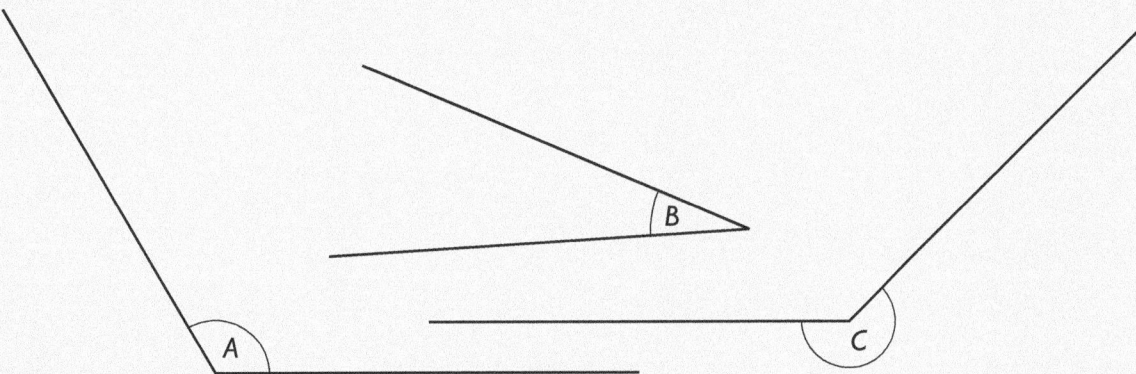

4. Draw and label the following angles.
   (a) 72°   (b) 107°   (c) 154°
   (d) 3°   (e) 200°   (f) 333°

5. For each of these diagrams (i) make an approximate copy
   (ii) calculate the missing angle   (iii) give your reason.

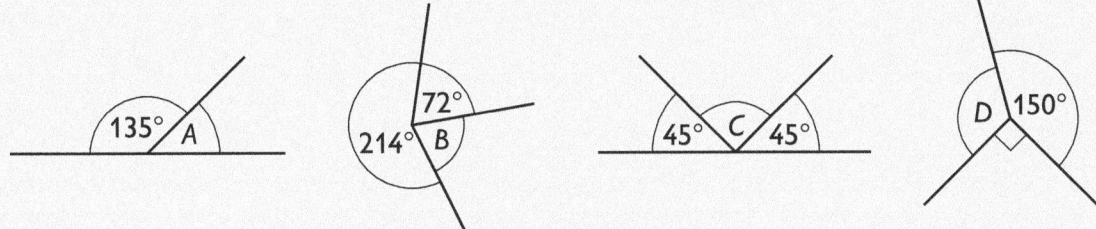

6. Calculate the value of each lettered angle.
   The diagrams are not drawn accurately.

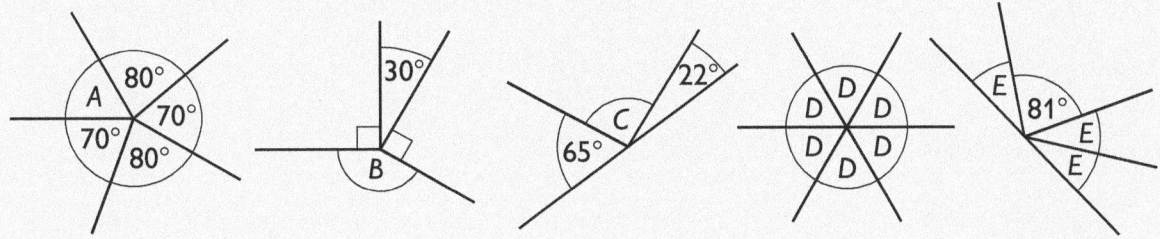

Angles at intersecting lines

 What can you say about the distance between two parallel lines?
Think of ten sets of parallel lines in your classroom.

Here is part of a wire fence going up a hill.

The wires are parallel.
The posts are vertical.

 Why must the posts be parallel?

Task

I Measure the angles in the picture and copy and complete these tables:

A	B	C	D

P	Q	R	S

W	X	Y	Z

 What do you notice about the angles?

2 Make a tracing where a wire intersects a post. Place it over the other intersections.
3 Rotate the tracing by half a turn. It is now upside down.
Place the upside down tracing over different intersections.

The Task shows you three facts about angles made by intersecting lines:

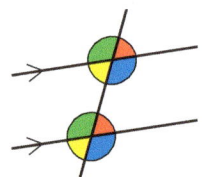

Corresponding angles
are equal

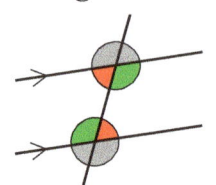

Alternate angles
are equal

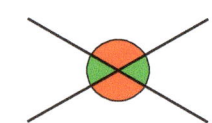

Vertically opposite angles
are equal

Exercise

1   Make an approximate copy of each of these diagrams.
    For each letter **(i)** work out the angle **(ii)** give your reason.

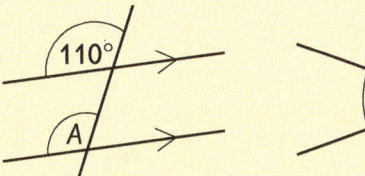 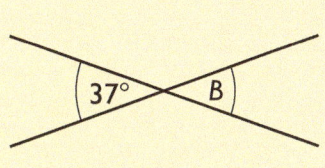 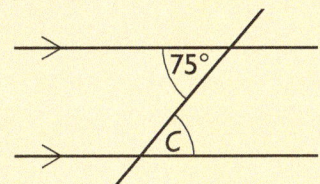

2   Calculate the value of each lettered angle.

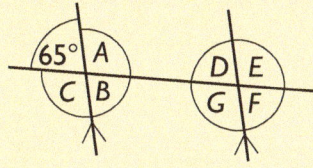 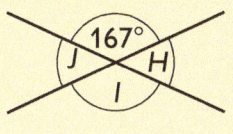 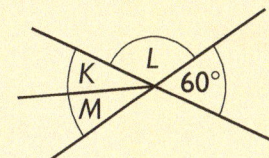

3   How many different sizes of angle are there
    in this diagram?

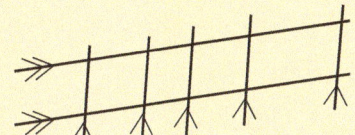

4   **(a)** Find the value of each lettered angle.

    (b) What is $A + B + C + D$?

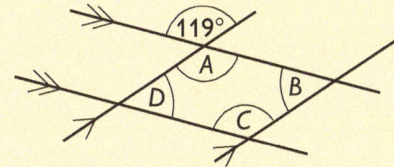

5   Write down the value of each letter.
    For each one give your reason.
    What is $F + D + E$?

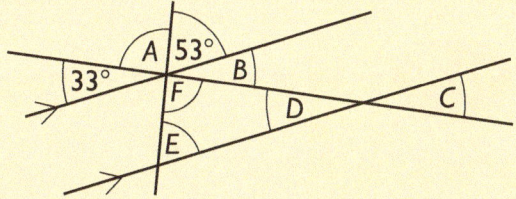

6   Here is a fence going down a hill.
    The wires are parallel. The posts are vertical.

    (a) What is angle A?

    (b) What angle do the wires make with the
    horizontal?

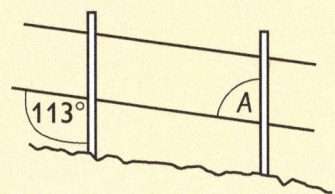

Investigation

Here are three pairs of intersecting parallel lines.

1   Sketch a copy of the diagram.

2   Find the size of each lettered angle.

3   What can you say about the triangles
    in the diagram?

4   Explain your answer.

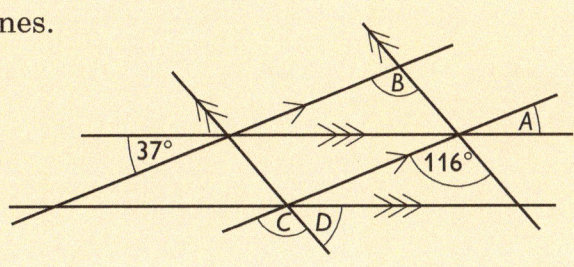

Angles in a triangle

The angles inside a shape are called **interior angles**.
An angle made by extending a side is called an **exterior angle**.

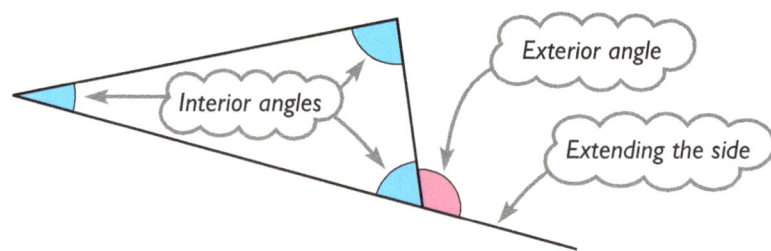

Exterior angle

Interior angles

Extending the side

Task

Look at this diagram.
The horizontal lines are parallel

1 Why is angle *B* 47°?
2 Why is angle *A* 72°?
3 Why is angle *C* 61°?
4 What is *A* + *B* + *C*?
5 What is the sum of the interior angles of the triangle?

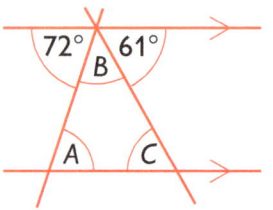

6 **(a)** Copy this diagram.
  (b) Explain why the sum of the interior angles of *any* triangle is 180°.
    Is your explanation a **proof**?

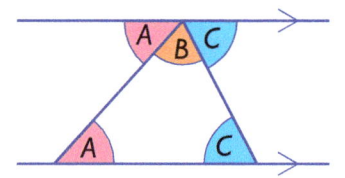

Task

Look at the diagram.

1 Why is angle *C* 32°?
2 Why is angle *D* 148°?
3 How is angle *D* related to 54° and 94°?

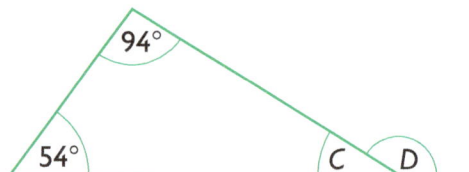

4 **(a)** Copy this diagram.
  (b) Show that an exterior angle of *any* triangle equals the sum of the opposite two interior angles.

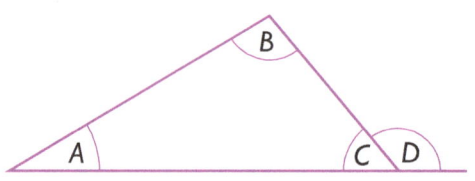

**There is another proof using parallel lines.
How does this diagram show that:
'The exterior angle of a triangle equals the sum of the opposite two interior angles.'
What is a proof?**

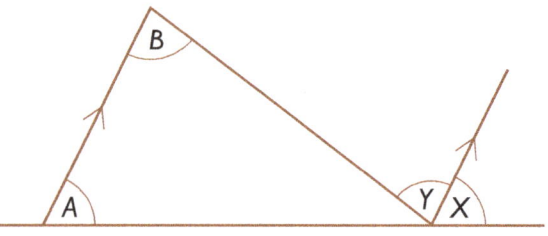

Exercise

1 **(a)** Make an approximate copy of each of these diagrams.

(b) Work out the size of each lettered angle. Give your reason.

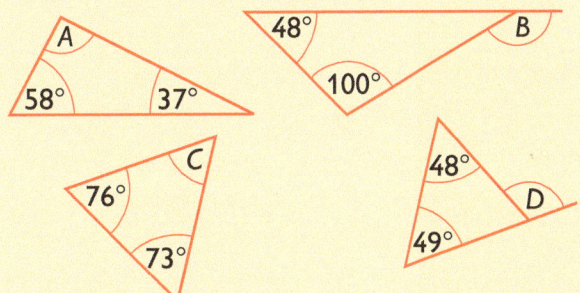

2 This quadrilateral is split into two triangles.

(a) Copy the diagram.

(b) What is $A + B + C$?

(c) What is $P + Q + R$?

(d) Explain why the sum of the interior angles of the quadrilateral is 360°.

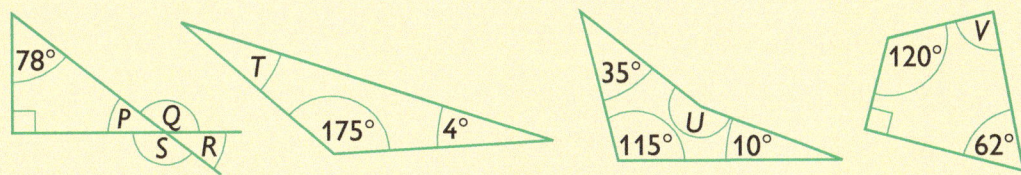

3 Calculate the size of each lettered angle in the following shapes.

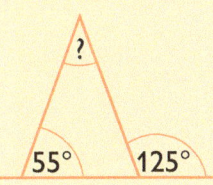

4 Calculate the acute angle between the wall and the ladder.

5 Calculate the angle at the apex of the tent.

6 Calculate the size of each lettered angle in the following shapes.

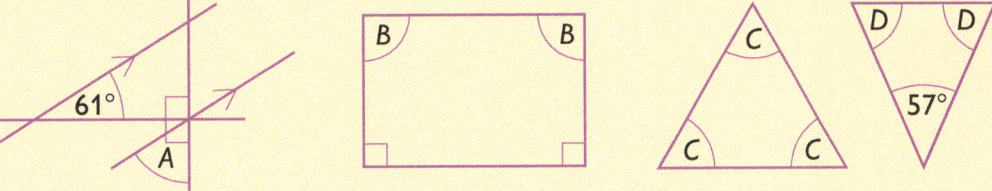

7 Here is the end wall of a house. What is angle X? What assumptions do you have to make to answer this question?

8 What is angle Y?

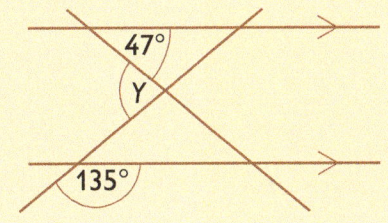

Bearings

A pilot is flying from Edinburgh to London.

 How does she check she is flying in the correct direction?

What *is* the correct direction?

How accurately can you describe the direction?

❝ Do the right thing!

Direction is described by a **compass bearing**.

This is measured clockwise from the north.

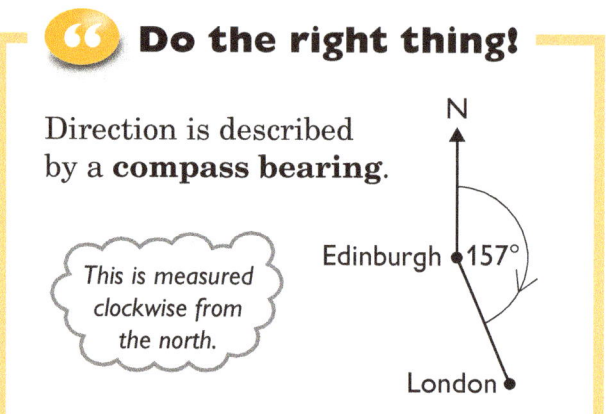

The bearing of London *from* Edinburgh is 157°.

Say 'one five seven degrees'

Check this on the map using a protractor.

Compass bearings are always written and said using three figures.

 What are the bearings of North, South, East, West, North-West and North-East?
What are the smallest and largest possible bearings?

Task

1 **(a)** The pilot's return journey *from* London to Edinburgh is a bearing of 337°.
This is a **back bearing**.
How is 157° related to 337°? How can you *calculate* a back bearing from a bearing?

(b) The bearing of Swansea from Carlisle is 191°.
Calculate the back bearing and check with a protractor.
What has happened in your calculation?

2 Measure the bearings of **(a)** Norwich from Swansea and **(b)** London from Penzance.
What can you say about these two journeys?

Exercise

1 Measure the following compass bearings:
- **(a)** Norwich from Penzance
- **(b)** Birmingham from Newcastle
- **(c)** Swansea from Carlisle
- **(d)** Swansea from Norwich
- **(e)** Carlisle from Newcastle
- **(f)** Inverness from Swansea
- **(g)** London from Newcastle
- **(h)** Norwich from Carlisle

2 **(a)** A pilot flies from Edinburgh to Newcastle on a bearing of 146°. Calculate the back bearing for the return journey?
(b) The flight path from Carlisle to Inverness is 345°. Calculate the bearing of Inverness from Carlisle.

3 Dave and Ann are walking on Kinder Scout in the Peak District. Their route is Seal Stones to Kinder Downfall, then to Noe Stool, then back to Seal Stones.
- **(a)** Calculate angle A.
- **(b)** Calculate the bearing of Kinder Downfall from Seal Stones.
- **(c)** Write down the bearing of Noe Stool from Kinder Downfall.
- **(d)** What bearing must they follow to return from Noe Stool to Seal Stones?

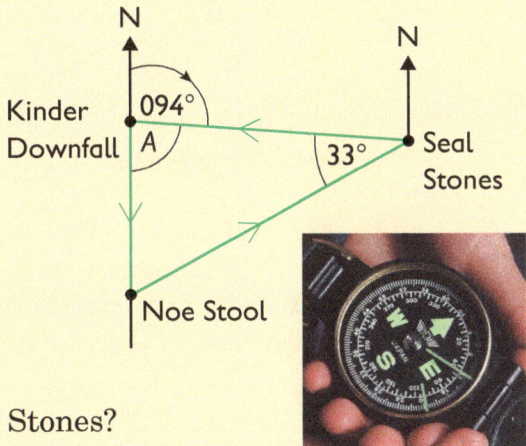
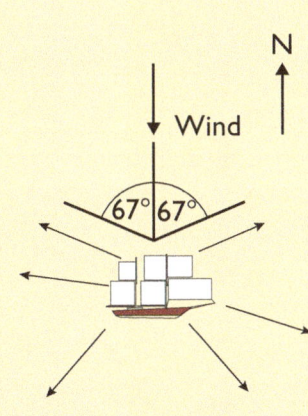

4 The bearing of Edinburgh from Inverness is 157°.
The bearing of London from Edinburgh is also 157°.
- **(a)** What can you say about the positions of these three places?
- **(b)** What is the bearing of London from Inverness?

5 Inverness, Carlisle and Birmingham all lie on the same straight line.
- **(a)** Measure the bearing of Birmingham from Carlisle.
- **(b)** From your answer to part (a), calculate the bearing of Inverness from Carlisle.

6 Sailing ships cannot sail directly into the wind.

However, they can sail towards the wind but only at an angle to its direction.

Captain Kipper's square rigger cannot sail closer than 67° into the wind.
It can sail in all other directions.
In the diagram, the wind is *from* the North.
It is a 'North wind'.

Between which directions can he *not* sail in
- **(a)** a North Wind
- **(b)** a South wind
- **(c)** an East wind
- **(d)** a South-West wind?

Finishing off

Review exercise

1 For each of the following diagrams
   (i)   make an approximate copy.
   (ii) calculate the size of the lettered angle(s).
   (iii) write down your reason(s).

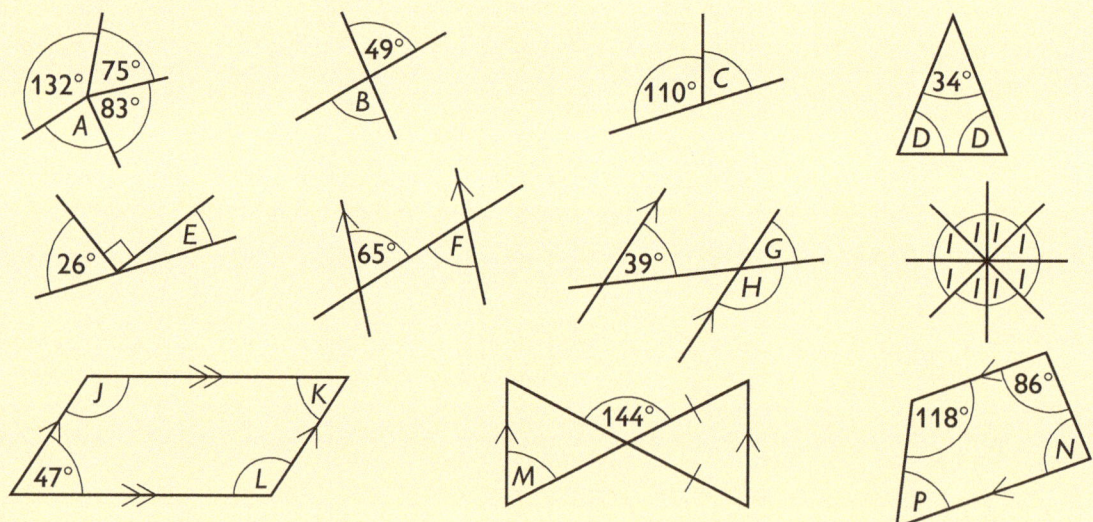

2 How many degrees does the minute hand of a clock turn through in 13 minutes?

3 (a) Draw and label angles of the following sizes:
      (i) 58°   (ii) 213°   (iii) 90°   (iv) 111°
      (v) 180°   (vi) 300°   (vii) 354°

   (b) Ask a friend to check the accuracy of your drawings.

   (c) Next to each angle in part (a) write down the type of angle.
       For example 58° is acute.

4 **(a)** Measure the bearings of the following journeys, shown on the map below.

    (i) Filwood to Durton   **(ii)** Durton to Borley

    (iii) Filwood to Egwell   **(iv)** Borley to Filwood

    (v) Carham to Filwood   **(vi)** Borley to Carham.

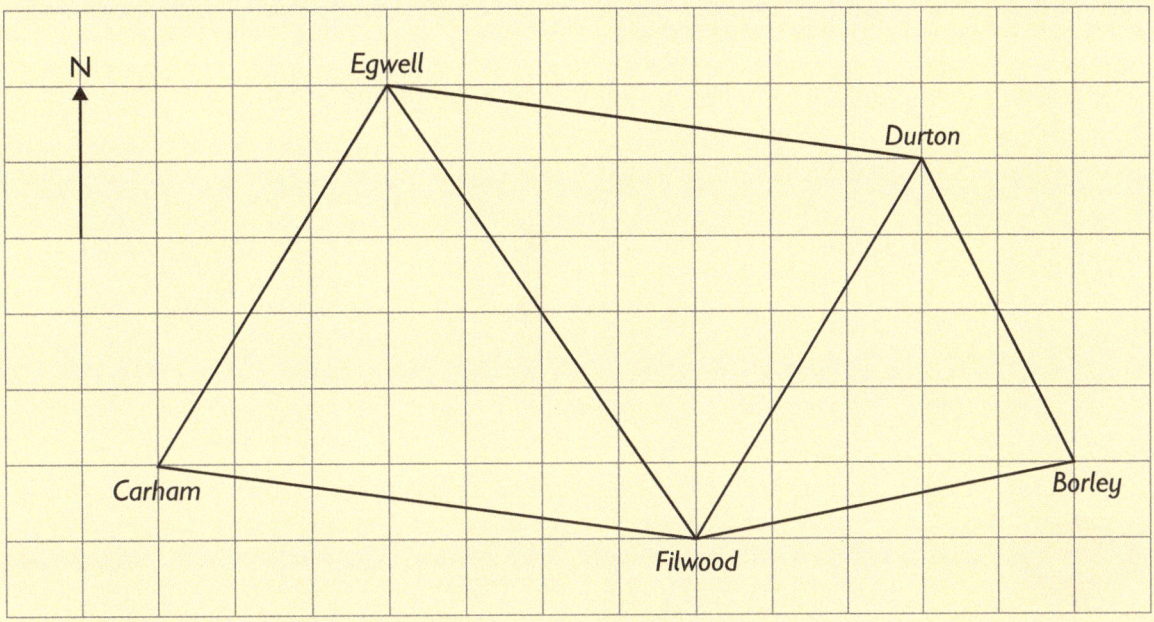

(b) Calculate the back bearing of each journey in part (a).

(c) Which journeys drawn on the map have the same bearings?

5 Calculate angles *A* and *B* in this electricity pylon.

6 Make a copy of this diagram, which is not drawn accurately.

Find angle *X*.

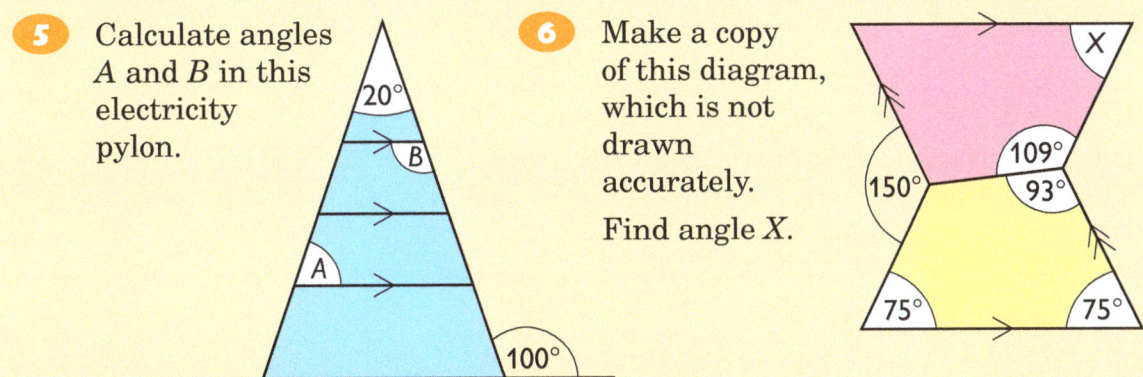

7 The bearing of Twigham from Manor is 081°. Calculate the bearing of Arbour from Manor.

8 Make a sketch copy of this pentagon. Prove that its interior angles add up to 540°.

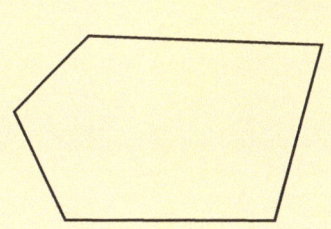

4 Displaying data

Julia records the results of her hockey club during a season.

She uses a **tally chart** to organise them.

She presents these results in other ways.

	Tally	Total
Win	ⵏⵏⵏⵏ ||||	9
Lose	ⵏⵏⵏ |	6
Draw	|||	3

Pictogram

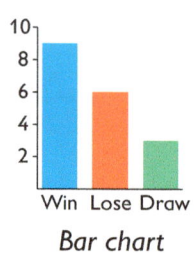

Bar chart

Pie chart

 How did she work out the angles to use in the pie chart?
What are the advantages and disadvantages of the three different displays?

Task

Show this picture to your classmates. Give them 10 seconds to remember as many objects as they can.

(a) Make a tally chart to record how many objects each can remember.

(b) Make a frequency table for the data.

(c) Draw a bar chart to illustrate your results.

 Are the data suitable for a pie chart?

Grouped data

Roshan keeps a record of his rugby team's scores during each match of the season.

```
0    26   15   32   35   18
18   19   21   14   16   16
29   14    3   25    9   19
 8   28   11   20   30   13
```

Task

Using the class intervals 0–9, 10–19, 20–29 and 30–39 for *Number of points scored*, construct a tally chart showing Roshan's data.

 Why does he group these data before he displays them?
How would you display Roshan's grouped data?

Exercise

1 The bar chart shows the number of days it rained during September in cities in England.

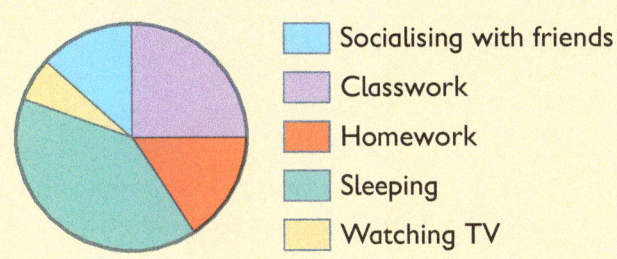

(a) Which city had the most days of rain?

(b) Draw a pictogram to illustrate these data. Use ☁ to represent 2 days.

2 This pie chart shows the way Sylvia spent one day.

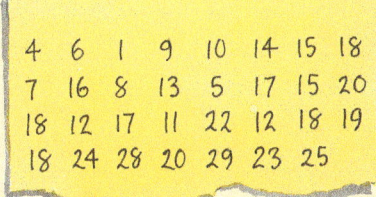

- Socialising with friends
- Classwork
- Homework
- Sleeping
- Watching TV

(a) What does she do for longer than anything else?

(b) Which activity takes up $\frac{1}{4}$ of her day?

(c) On which two activities does she spend the same amount of time?

3 Ronny goes to the City of London. He records the number of floors of the buildings within 100 metres of Liverpool Street Station.

Here are his results.

```
4   6   1   9   10  14  15  18
7   16  8   13  5   17  15  20
18  12  17  11  22  12  18  19
18  24  28  20  29  23  25
```

(a) How many buildings are within 100 metres of Liverpool Street Station?

(b) Make a tally chart using groups 1–5, 6–10 and so on.

(c) Give the grouped data as a frequency table.

(d) Draw a bar chart to illustrate the grouped data.

(e) Which is the modal class?

Activity In the Task on the page opposite you made a tally chart of the scores for Roshan's rugby team. Here is another way to show the scores. It is a **stem-and-leaf** diagram.

Stem	Leaves	key 1|5 represents 15
0	0 3 9 8	
1	5 8 8 9 4 6 6 4 9 1 3	
2	6 1 9 5 8 0	
3	2 5 0	

This score is 10 + 1 = 11

This score is 20 + 8 = 28

(a) Rewrite this stem-and-leaf diagram with the leaves written in order.

(b) Make a stem-and-leaf diagram for Ronny's data in question 3.

(c) How is a stem-and-leaf diagram better than a tally?

The last line becomes
0   2   5

Mean, median and mode

Three friends are discussing whether this is true.

That can't be right – my dad takes size 39.

Fred

I think its an April Fool.

Giovanna

I take size 43 already. I reckon it's true.

Gene

They ask the men who walk past their school during a 15 minute period 'What size shoes do you take?'

Here are their results.

41	40	43	40
41	43	47	40
43	43	41

 **Task**

Survey the people in your class to find out their shoe size.

They each calculate an average.

Fred calculates the mean:

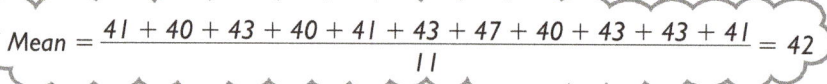

$$\text{Mean} = \frac{41 + 40 + 43 + 40 + 41 + 43 + 47 + 40 + 43 + 43 + 41}{11} = 42$$

Giovanna finds the median:

The median is the middle value when all the data are arranged in order

40 40 40 41 41 (41) 43 43 43 43 47

Median

Gene finds the mode:

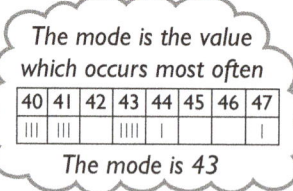

The mode is the value which occurs most often

40	41	42	43	44	45	46	47
III	III		IIII	I			I

The mode is 43

? **They are all right, but who has chosen the best average?**

? **How would Giovanna have found the median if there had been an even number of data values?**

 **Task**

Calculate the mean, median and mode for the shoe size of the people in your class.

Exercise

1 For the following numbers find
    (i) the mean   **(ii)** the median   **(iii)** the mode.

    (a) 3, 4, 5, 9, 9

> To find the median first arrange the numbers in order of size

    (b) 8, 10, 10, 10, 17

    (c) 12, 2, 1, 1, 3, 32, 1, 2, 4, 2

    (d) 45, 62, 24, 45, 19, 82, 16, 7, 16, 8, 16, 8

    (e) 6, 2, 4, 10, 20, 18, 6, 9, 12, 2, 1, 5, 2, 10, 13

2 For the following state whether the mean, median or mode is the best average to use. Explain your answers.

    (a) A cricketer's average score for the season's matches.

    (b) The size of helmet a school cricket team buys to share.

    (c) An employee of a record company wants to know the average salary of people in her company. They all tell her their salaries including the director.

3 Jack and Delroy are comparing their results in the end of year examinations.

Jack's exam results (%)								
61	67	70	63	48	60	40	42	44
Delroy's exam results (%)								
75	59	82	94	55	80	53	48	48

    (a) Find the median score for each boy.

    (b) Find the mean score for each boy.

    (c) Who do you think has performed better?

Activity   Fill a jam jar with sweets.

Ask everyone in the class to estimate the total number of sweets.

Write down the answers.

Find the mean, median and mode of your results.

Whose value is nearest to the correct answer?

 Find the averages for the boys and for the girls.
Are the boys or girls in your class better at estimating?

Measuring the spread of data: range

Ginger and her friends now ask 11 women about their shoe sizes. Here are their results.

They decide to find out the spread of the data.

They calculate the *range*.

Range = highest data value − lowest data value
= 42 − 36 = 6

Task

Calculate the mean, median and mode for the women's shoe sizes.
Compare the data for men and women.
Which of the two sets has the greater range?

Outliers

Outliers are data values which do not fit in with the rest of the data.

Jan asks the age of all the competitors in two local BMX competitions.

BMX Extreme Games
15 12 13 11 20 14 58

Superbike Championships
16 15 11 144 13 15 14 13 15 16

? **Which value is an outlier in each case?**

Task

Remove the mistake. Now calculate the range of each data set.
Which of the two BMX competitions has the larger spread of age?

⚠ Only remove an outlier if you are *sure* that it is an error!

Exercise

1 All these data sets include outliers.

(i) Audrey's examination marks (%) 58, 26, 74, 29, 51, 154, 68

(ii) Age of competitors at a freestyle skateboard contest 15, 12, 18, 13, 54, 17, 12

(iii) Total rolled on two dice 3, 10, 5 , 6, 16, 7, 9, 2, 8, 7

(iv) Number of kilometres run by Gary on each day of a certain week 5, 8, 5, 7, 4, 2, 42

In each case:

(a) State the value of the outlier.

(b) Decide whether you should remove it. Explain why.

(c) Calculate the range.

2 Caroline is taking a typing test.
She types for 1 minute and the number of correct words is recorded.
She does this 6 times.

46   49   44   43   45   49

In order to pass she must have
● a mean of greater than 45.
● a range of less than 7.

Does she pass?

3 Here are the lap times for two racing drivers during a five lap race.

Andy Stephen					Kevin Henderson				
64 s	66 s	63 s	63 s	64 s	66 s	67 s	61 s	66 s	60 s

(a) Calculate the mean and range for both drivers.

(b) Who drives the fastest lap?

(c) Who drives the slowest lap?

(d) Who is the more consistent driver?

(e) Who wins the race?

Activity   Imagine you are commentating on the race in Question 3.
Write down what you say. Think particularly about
● their positions after each lap
● who starts faster?
● who finishes faster?
Before the race the lap record was 61 s. Has this changed?

Discrete and continuous data

Hilary and Todd are collecting data.

How many pets have you got?

0	2	3	0	1
4	2	0	1	6
5	0	3	2	2
1	0	0	3	4
2	4	3	1	1

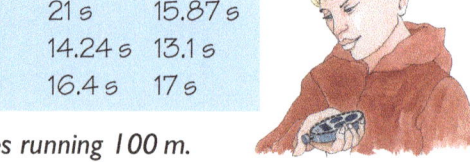

14 secs	13.5 s	20 s	16.7 s	14.96 s
15 s	19 s	16.75 s	14.8 s	17.63 s
13.9 s	17.2 s	18 s	21 s	15.87 s
18.2 s	17.3 s	20 s	14.24 s	13.1 s
16 s	18.12 s	14 s	16.4 s	17 s

Todd times his classmates running 100 m.

 What is the difference between Hilary's data and Todd's data?

For Hilary's data each value must be a whole number.
You cannot own 4.2 pets! This type of data is called **discrete** data.

Todd's data can take any value.
The time taken to complete 100 metres could be 14 seconds, 15.2 seconds, 13.98 seconds, etc. This type of data is called **continuous** data.

Write down **(a)** 5 examples of discrete data and **(b)** 5 examples of continuous data.

Todd wishes to *group* his data using a frequency table. He could use either of these two options.

Table 1

Time (secs)		
At least	Below	Frequency
12	14	
14	16	
16	18	
18	20	
20	22	

Table 2

Time, t (secs)	Frequency
$12 \leqslant t < 14$	
$14 \leqslant t < 16$	
$16 \leqslant t < 18$	
$18 \leqslant t < 20$	
$20 \leqslant t < 22$	

 The first time is 14 s. Does it belong to the first or the second group in Table 1. What about in Table 2?

Why can Todd not use the groups 12–14, 14–16, 16–18, …?
Why can he not use the groups 12–13, 14–15, 16–17, …?

Copy and complete Tables 1 and 2.

Are there other ways Todd can group his data?

Exercise

1 Are the following data discrete or continuous?

(a) The number of people in cars passing the front of your school.

(b) The amount of money earned doing Saturday jobs.

(c) The wing-span of eagles.

(d) The score showing on a die when rolled.

(e) The height a ball reaches when bounced.

(f) The weight of new born dolphins.

2 The maximum temperature, in °C, is recorded in Auckland for each day in June.

16.4	12.8	17.6	19.1	16.6	15.5
11.2	18.7	19.5	16.1	15.3	14.2
15.8	15.7	14.9	14.4	13.4	12.1
13.9	11.9	13.1	12.6	10.9	13.5
14.2	15.4	16.6	15.9	15.6	14.3

June

(a) Copy and complete this tally chart and frequency table.

Temperature (°C)	Tally	Frequency
$10 \leqslant$ Temp. < 12		
$12 \leqslant$ Temp. < 14		
$14 \leqslant$ Temp. < 16		
$16 \leqslant$ Temp. < 18		
$18 \leqslant$ Temp. < 20		

On how many days was the temperature

(b) at least 16°C but less than 18°C

(c) less than 14°C   **(d)** at least 12°C?

3 The time, in seconds, for 25 fireworks to burn out was recorded.

10	21	21	27	20
16	27	31	14	18
29	15	32	35	33
21	26	22	19	29
25	20	22	28	23

Using the class intervals $10 \leqslant t < 15$, $15 \leqslant t < 20$ and so on, draw and complete a tally chart and frequency table.

Activity

1 Write the numbers 1–10 randomly as shown opposite.

2 Make a number of copies.

3 Use a stopwatch to time how long your classmates take to join up the numbers in ascending order.

4 Make a tally chart and frequency table showing your results.

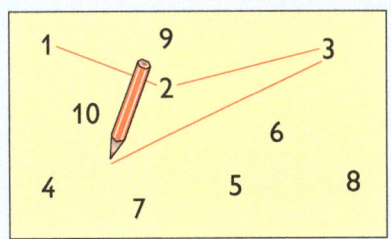

Scatter diagrams

Tim and Vicky are having an argument.

I think that people with long legs can jump further than people with short legs.

Rubbish! The length of a person's legs does not affect how far they can jump.

They decide to collect data from their friends to find out who is right.

	Alan	Barry	Claire	Dipak	Ernie	Flora	Gerry	Humza	Ian
Inside leg measurement (cm)	60	70	50	65	65	70	60	75	55
Standing jump distance (cm)	85	90	65	90	80	100	70	95	80

Task

1 Draw a **scatter diagram** by plotting the points (60, 85), (70, 90), (50, 65) and so on.

2 The scatter diagram has been started for you. Copy and complete.

? **Does the scatter diagram support Tim or Vicky?**
Have they sampled enough people to be certain?

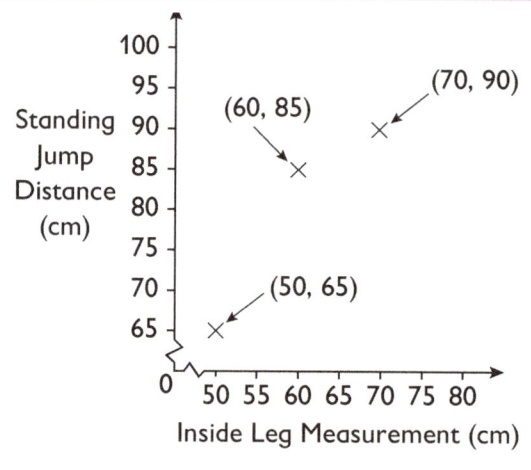

The word **correlation** describes the relationship between the values of the two variables.

If the two values *increase* together, there is **positive correlation**.

If one value *decreases* as the other value *increases*, there is **negative correlation**.

If there is no relationship between the values, there is **no correlation**.

 Among Tim and Vicky's friends what type of correlation exists between the length of a person's legs and how far they can jump?

 Think of one everyday example of (a) positive correlation (b) negative correlation (c) no correlation.

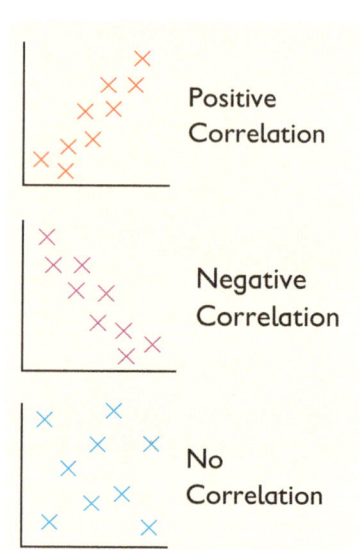

Positive Correlation

Negative Correlation

No Correlation

Exercise

1   Here are some examples of scatter diagrams. Fill in the blanks in the following sentences. Use only the words *increases, decreases, positive, negative* and *no*.

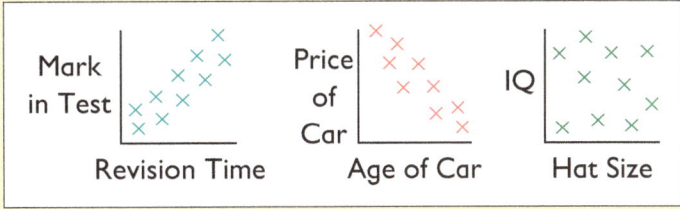

(a) As the *revision time* _____ the *test mark* _____, so there is _____ correlation between revision time and test mark.

(b) As the *age of car* _____ the *price* _____, so there is _____ correlation between age and price of car.

(c) Hat size does not appear to be related to IQ, so there is _____ correlation between hat size and IQ.

2   This table shows the marks (out of 10) given by judges at a local vegetable show.

	Mrs Giles	Mr Hands	Mr Smith	Mr Taylor	Mr Thomas	Ms Barrett	Mrs Hogg	Mr Elphick	Mr Wade
Judge 1	1	6	3	5	7	2	8	9	4
Judge 2	2	8	3	5	8	3	10	7	6

(a) Draw a scatter diagram showing the marks of the two judges.

(b) Is there any correlation between the marks of the two judges. If so, what kind?

(c) Who was the only person to get a higher mark from judge 1 than judge 2?

(d) The winner was the person with the highest overall mark. Who won the competition?

3   Trevor measures the floor area of a number of classrooms in his school. He also counts the number of desks in them.

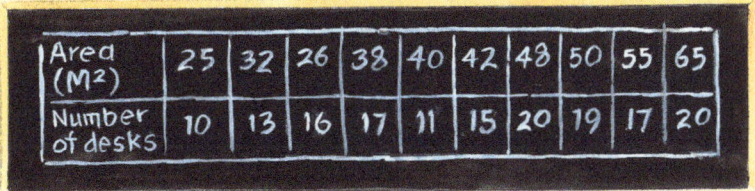

Area (m²)	25	32	26	38	40	42	48	50	55	65
Number of desks	10	13	16	17	11	15	20	19	17	20

(a) One classroom has area 50 m². How many desks does it have?

(b) What is the floor area of the classroom with 16 desks in it?

(c) Draw a scatter diagram.

(d) Another classroom has floor area 46 m². Use your scatter diagram to estimate how many desks it can have.

(e) Comment on the correlation between the area and the number of desks.

Finishing off

Now that you have finished this chapter you should be able to:

- represent data using a pictogram, bar chart or pie chart
- find the mean, median and mode, and know when to use them
- find the range and understand that it is a measure of how spread out the data are
- understand the difference between discrete and continuous data, and be able to group both types using a frequency table
- know when to use a scatter diagram and understand the meaning of correlation.

Review exercise

1 Shamicka asks her schoolfriends to record the number of minutes they spend playing computer games during May.

Here are her results.

25	150	262	30	143	0	55	320	260	60
65	140	40	170	74	130	45	125	300	220
96	132	90	185	89	167	68	50	160	82

(a) Copy and complete this tally chart.

Time spent playing computer games (min)	Tally	Frequency
0–50		
51–100		

(b) Draw a bar chart to illustrate these data.
Shamicka plans to ask the same question in August.
(c) Predict the shape of the new bar chart.

2 The members of a computer games club are asked to give a mark out of 10 for the latest game *Crypt Stormer*.

Boys	6	7	3	4	3	6	3	8
Girls	3	5	77	2	7	8	5	–

(a) Which value is an outlier? Do you include it?
(b) Calculate the median and mean for the boys and girls.
Who rate the games higher, the boys or the girls?
(c) Calculate the range for the boys and girls.
Whose marks are more spread out?

3 Are the following data discrete or continuous?
(a) The points scored by a group of friends on a computer game.
(b) The numbers of pupils in each class at school.
(c) The weights of moon rock specimens.
(d) The distances travelled by snails in a day.
(e) The numbers of letters arriving each day at your house.

4 Jules asks her friends to sing a musical note for as long as possible without taking a breath. She records her results in a frequency table.

Time (secs)		Frequency
At least	**Below**	
10	15	3
15	20	6
20	25	4
25	30	3
30	35	2

(a) How many people take part in the experiment?

(b) What is the smallest amount of time someone in the class may have sung the note?

Jules's friend Paul arrives late and is not included in the original table. He sings a note for exactly 35 seconds.

(c) Paul's result cannot be recorded in the table. Explain why.

(d) Rewrite the table using the groups $10 \leqslant t < 20, 20 \leqslant t < 30, 30 \leqslant t \leqslant 40$. Include Paul's result. What does t stand for?

5 Look at the scatter diagram opposite. It shows the weight and cost of 4 books taken off a shelf at an antique book shop.

Julie says 'You can see that heavier books cost more'.

(a) Why does Julie say this?

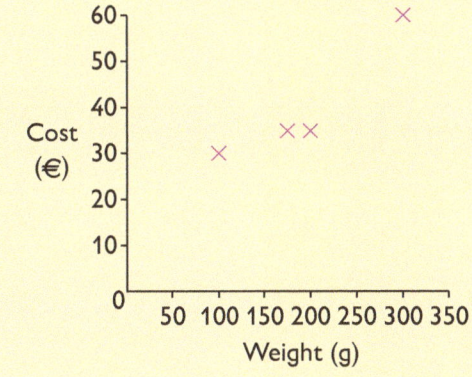

Another 7 books are taken off the shelf at random.

Weight (g)	100	250	55	200	205	300	50
Cost (€)	45	15	50	8	10	7	36

(b) Copy the scatter diagram above, and add the new values.

Robert says 'There is no correlation between a book's weight and its cost'.

(c) Which of the two opinions is the more reliable, Julie's or Robert's?

6 A group of friends take part in a sponsored swim.

Name	Mohammed	Sonia	Elizabeth	Malcolm	Angela	Carrie	Michelle
Age (years)	20	38	60	51	28	33	26
No. of lengths	30	19	14	11	26	34	42

(a) Draw a scatter diagram showing these data.

(b) What type of correlation exists between the ages of these swimmers and the number of lengths they swim?

(c) Terry (age 46) and Clara (age 31) also swim. Can you say who swims further?

5 **Decimals**

Rounding and decimal places

Colin is trying to place the number 35.654 on a number line.
First he looks at the whole numbers.

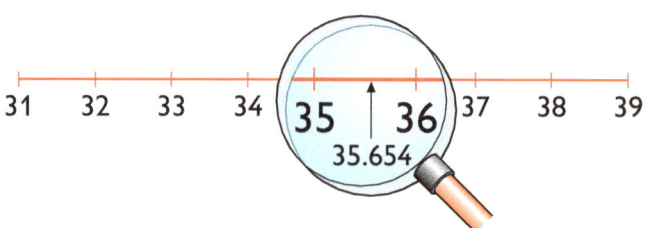

*35.654 lies between 35 and 36.
It is nearer to 36 than 35.*

? **What is 35.654 to the nearest whole number?**
Which of the numbers after the decimal point helps you decide this?

Next he looks at the first decimal place.

35.654 lies between 35.6 and 35.7.

? **Colin has marked the number on the line. Do you agree with him?**
What is 35.654 to 1 decimal place?
Explain why you had to look at both the 2nd and 3rd decimal places to help you decide this.

```
|——+——+——+——+——+——+——+——|
35.61  35.62  35.63  35.64  35.65  35.66  35.67  35.68
```

35.654 lies between 35.65 and 35.66.

? **Use this to place the number in the correct place on a copy of the number line above?**
What is 35.654 to 2 decimal places?

Task

1 Place the number 8.156 on an accurate number line.
2 Round 8.156 **(a)** to the nearest whole number **(b)** to one decimal place
(c) to two decimal places

? **Colin must write 5.364 to 1 decimal place.**
He chooses between 5.3 and 5.4. Which should he choose? Explain why.

? **11.49 < 11.498 < 11.50** **Write 11.498 to 2 decimal places.**

Exercise

1   Write down your steps for the following:

   (a) Start at 3 and count in intervals of 0.1 until you reach 4.

   (b) Count in intervals of 0.1 from 6.5 to 7.5.

   (c) Start at 3 and count in intervals of 0.01 until you reach 3.1.

   (d) Count in intervals of 0.01 from 4.95 to 5.05.

2   **(a)** Add 0.1 to each of the following:
     (i) 4   **(ii)** 7.6   **(iii)** 10   **(iv)** 2.9   **(v)** 9.9

   (b) Subtract 0.1 from each of the following:
     (i) 8.4   **(ii)** 0.6   **(iii)** 0.12   **(iv)** 10   **(v)** 100

   (c) Write down the number that is 0.01 more than:
     (i) 6.48   **(ii)** 12.33   **(iii)** 0.02   **(iv)** 0.09   **(v)** 9.99

   (d) Write down the number that is 0.01 less than:
     (i) 7.56   **(ii)** 0.04   **(iii)** 3.452   **(iv)** 0.1   **(v)** 1

3   72.386 lies between 72.3 and 72.4.

   72.386 = 72.4 correct to 1 decimal place.
   For each number below **(i)** round to 1 decimal place
                              (ii) write two sentences like the ones above to
                                    explain your choice.

   (a) 13.243   **(b)** 0.196   **(c)** 6.091

   (d) 0.3257   **(e)** 0.06   **(f)** 0.09

4   Write each of the following numbers **(i)** to 1 decimal place
                                        (ii) to 2 decimal places.

   (a) 12.6723   **(b)** 1.438   **(c)** 0.6951   **(d)** 0.076

   (e) 1.0359   **(f)** 4.0032   **(g)** 9.951   **(h)** 0.9999

5   Write down the smaller number from each of these pairs.

   (a) 3.56 and 3.471   **(b)** 8.95 and 8.899   **(c)** 1.334 and 1.34

   (d) 0.12 and 0.121   **(e)** 0.2256 and 0.227   **(f)** 0.015 63 and 0.015 512

6   Find the number that lies half way between:

   (a) 2.95 and 2.96   **(b)** 3.99 and 4.01   **(c)** 0.23 and 0.37

7   Do the following calculations on your calculator.
   Give your answers to 2 decimal places.

   (a) 5 ÷ 3   **(b)** 10 ÷ 6

   (c) 14 ÷ 9   **(d)** 23 ÷ 7

Rounding errors

 What is the cost of one complete hot dog with tomato ketchup?

Richard works out the cost.

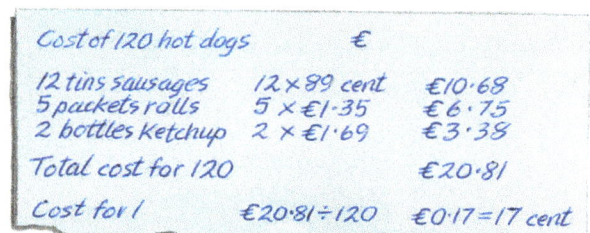

Cost of 120 hot dogs		€
12 tins sausages	12 × 89 cent	€10·68
5 packets rolls	5 × €1·35	€6·75
2 bottles Ketchup	2 × €1·69	€3·38
Total cost for 120		€20·81
Cost for 1	€20·81÷120	€0·17 = 17 cent

Anna works out the cost.

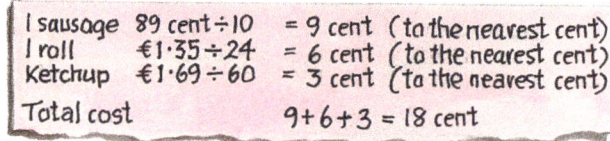

1 sausage	89 cent ÷ 10	= 9 cent (to the nearest cent)
1 roll	€1·35 ÷ 24	= 6 cent (to the nearest cent)
Ketchup	€1·69 ÷ 60	= 3 cent (to the nearest cent)
Total cost		9 + 6 + 3 = 18 cent

 Explain why their answers are different. Which is the more accurate answer?

Freda works out this sequence on her calculator:

| 89 | ÷ | 10 | + | 135 | ÷ | 24 | + | 169 | ÷ | 60 | = |

 Work out the sequence on your calculator.
Why is this method more accurate than Anna's?

Most divisions do not give whole number answers.
When a calculation involves lots of dividing, leave the answer to each *part* in your calculator.
This gives a more accurate *final* answer.

 **Task**

1 Calculate 7 ÷ 3 and 14 ÷ 3 to 2 decimal places.

2 Use your answers to work out:
   (a) 7 ÷ 3 + 14 ÷ 3   **(b)** 14 ÷ 3 − 7 ÷ 3
   (c) 7 ÷ 3 × 14 ÷ 3   **(d)** (14 ÷ 3) ÷ (7 ÷ 3)

Ask your teacher to show you the brackets on your calculator.

 3 Do the same calculations again.
   This time work out the full calculation on your calculator before rounding the answers.

Look carefully at your answers to 2 and 3.

 Which answers are more accurate?
When have you got the same answer using both methods?
Work out 24 ÷ 7 × 14 ÷ 3 as accurately as possible.

Exercise

1 Work out the accurate cost of 1 patio pot containing 5 plants and 2 litres of compost.

€1·99 for 12   3 for €6·99   10 litres €5·99

2 Work out the following.
Give your answer to 2 decimal places when necessary.

(a) $(10 \div 3) + (5 \div 3)$ (b) $15 \div 7 \times 14 \div 3$ (c) $11 \div 7 + 16 \div 3$

(d) $8 \div 7 \times 3 \div 4$ (e) $7 \div 3 + 12 \div 7 + 10 \div 9 + 12 \div 11$

3 Work out the following.
Give your answers to the nearest cent.

(a) 20% of €5.99   (b) 20% of €2.99   (c) 20% of €1.59

(d) 20% of (€5.99 + €2.99 + €1.59)

(e) Explain why your answer to (d) cannot be found by adding together your answers to (a), (b), and (c).

Activity

George's calculator shows the number 7.
He does the following sequence of operations:

7

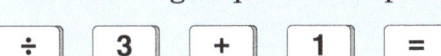

÷   3   +   1   =

1 What must he do to reverse this and get back to a display of 7?

Another time he uses the sequence

÷   8   ×   5   =

2 Write down the sequence that will reverse this.
Use your calculator to check that your answer works.

3 Choose a different number.
Work out   (your number + 7) ÷ 6
Reverse this to get back to your chosen number.
How would you reverse   (your number − 12) ÷ 11   ?
Check your answer on your calculator.

Investigation

For most numbers you cannot write the square root as an exact decimal. It goes on forever.

1 Write $\sqrt{18}$ correct to 2 decimal places.

2 Write $\sqrt{2}$ to 2 decimal places.

3 Use your answers to 1 and 2 to work out:

(a) $\sqrt{18} \times \sqrt{2}$ (b) $\sqrt{18} \div \sqrt{2}$ (c) $\sqrt{18} + \sqrt{2}$ (d) $\sqrt{18} - \sqrt{2}$

4 Do the same calculations again.
This time work out the full calculation on your calculator before rounding the answers.
Compare your answers.
Which of the operations $+ \; - \; \times$ or $\div$ gives the largest errors?

Interpreting decimal answers

? How long does it take to walk 14 km at a speed of 5 km per hour?

BOXLEY 14 km

? Five T-shirts cost €14. How much is this each?

Special offer

T-SHIRTS €3·50 each
OR   5 for €14

? 14 kg of sand is used to fill 5 sacks. What weight of sand is in each sack?

The problems above all involve the calculation $14 \div 5 = 2.8$
For each the decimal part (0.8) must be changed into an appropriate unit.

0.8 hours $= 0.8 \times 60$ minutes   *60 minutes = 1 hour.*

€$0.8 = 0.8 \times 100$ cent   *100 cent = €1.*

0.8 kilograms $= 0.8 \times 1000$ grams   *1000 grams = 1 kilogram.*

Work out the calculations above to obtain an appropriate answer to each problem.

? How many centimetres is 0.8 metres?

Task

George has worked out $123 \div 20$ on his calculator.

This is what his calculator shows: 6.15

1   The calculation involved time. Write the answer in hours and minutes.
2   The answer is in kilometres. Write it in km and metres.
3   The answer is in €. Write it in € and cent.
4   Write the answer in metres and centimetres.

You may wish to display mixed units as a decimal on your calculator.
The following sequence will change 3 hours 24 minutes into 3.4 hours.

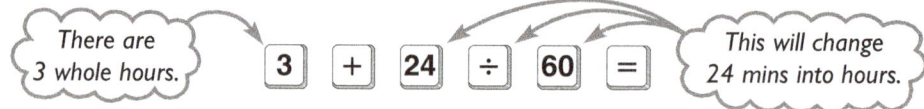

There are 3 whole hours.   3 + 24 ÷ 60 =   *This will change 24 mins into hours.*

? How would you change 5 hours 45 minutes into hours?

Exercise

1 Give an appropriate answer to the following problems:

   (a) Rashide cycles 45 km at an average speed of 11 km per hour.
How long does this take?

   (b) Vera is paid €35 for working 9 hours.
How much is this per hour?

   (c) Weldon takes 20 strides to walk 18 metres.
How long is each stride?

   (d) Alex collected €115 after doing a 35-km sponsored walk.
How much money is this per kilometre?

   (e) A lift can carry a maximum of 8 people or 650 kilograms.
What does this give as the average weight of a person?

   (f) A garden bench 2 m long seats 3 people.
How much room does each person have?

2 Write

   (a) €6.7 in € and cent.   **(b)** 21.47 km in km and metres.

   (c) 7.4 hours in hours and minutes.   **(d)** 8.2 hours in hours and minutes.

3 Write

   (a) €34 and 20 cent in €.   **(b)** 4 minutes 15 seconds in minutes.

   (c) 24 km and 45 m in km.   **(d)** 5 hours 12 minutes in hours.

   (e) 2 days 18 hours in days.   **(f)** 8 days 6 hours in days.

4 Find the cost of the following telephone calls:

   (a) 5 minutes 12 seconds at the weekend.

   (b) 3 minutes 25 seconds at peak rate.

   (c) 10 minutes 17 seconds off peak.

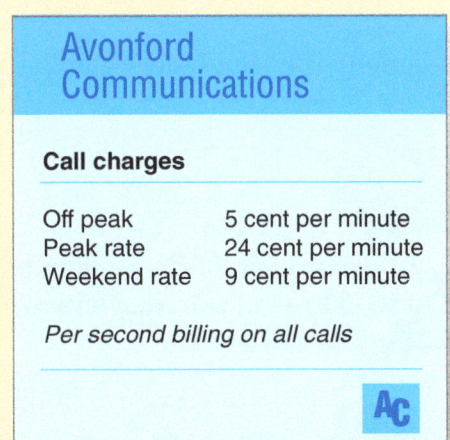

Avonford Communications

Call charges

Off peak	5 cent per minute
Peak rate	24 cent per minute
Weekend rate	9 cent per minute

Per second billing on all calls

Ac

5 Work out Jane's earnings for the week.

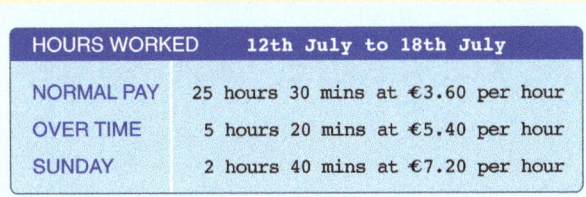

HOURS WORKED	12th July to 18th July
NORMAL PAY	25 hours 30 mins at €3.60 per hour
OVER TIME	5 hours 20 mins at €5.40 per hour
SUNDAY	2 hours 40 mins at €7.20 per hour

Rough calculations

Brad and his mother are arguing about the volume of this box.

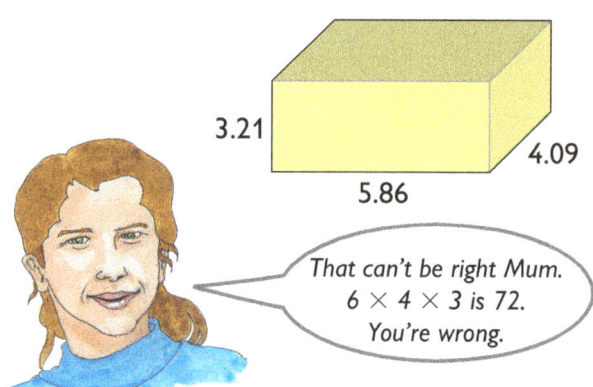

It's 19.9889

It must be right. My calculator says so.

That can't be right Mum. 6 × 4 × 3 is 72. You're wrong.

 How does Brad get 6 × 4 × 3?
Who is right, Brad or his mother?

It is always good to do a rough check against your calculator.
You may have pressed a wrong button.

Start by rounding the numbers in your calculation.

Example $\dfrac{28.4 \times 0.427}{4.92}$

Roughly $\dfrac{30 \times 0.4}{5} = 2.4$

Number	Rounded
28.4	30
0.427	0.4
4.92	5

A calculator gives the real answer, 2.4647 …

This is near the rough check of 2.4

 *OK.* ✓

The … means there are more numbers.

Task

For each of these calculations, do a rough check.
Three of the answers are wrong.
Which are they?

1. $26.43 \times 11.86 = 38.29$

2. $3.2 \times 4.1 \times 16.2 = 212.544$

3. $\dfrac{29.2 \times 161}{20.22} = 232.5 \ldots$

4. $5.1^2 = 10.2$

5. $62.3 + 317.9 - 12.43 = 87.875 \ldots$

6. $\dfrac{12.2 + 19.3}{0.41} = 76.829 \ldots$

 Each wrong answer in the Task comes from pressing a wrong calculator button.
Find the wrong button used in each case.

Exercise

1 These numbers appear in calculations. Round them for a rough check.
- **(a)** 9.0123
- **(b)** 212.41
- **(c)** 62.9123
- **(d)** 3.141 59
- **(e)** 1.006
- **(f)** 419
- **(g)** 6730
- **(h)** 2 146 309
- **(i)** 52.40
- **(j)** 0.102
- **(k)** 0.00812
- **(l)** 0.000 406
- **(m)** 0.0792
- **(n)** 0.987
- **(o)** 9.862

2 Des does some rough checks for his calculator sums.
For each one, answer either 'Near enough' or 'Too far away'.
- **(a)** Rough 6   Calculator 7.162 …
- **(b)** Rough 0.1   Calculator 10.29 …
- **(c)** Rough 450   Calculator 489.2 …
- **(d)** Rough 24   Calculator 2.414 …
- **(e)** Rough 0.048   Calculator 0.0511 …
- **(f)** Rough 17.5   Calculator 84.23 …

3 For each of the following
- **(i)**  do a rough check
- **(ii)** 📱 do the calculation on your calculator
- **(iii)** write down the right answer.

- **(a)** $8.6231 + 10.0019 - 5.832\,41$
- **(b)** 23.4×18.2
- **(c)** $79.6 \times 11.4 \times 102.6$
- **(d)** $\dfrac{41.8}{2.03}$
- **(e)** $\dfrac{3.921}{8.021}$
- **(f)** $\dfrac{28.9 \times 2.13}{42.6}$
- **(g)** $\dfrac{1238}{297}$
- **(h)** 0.029×389
- **(i)** $275.8 \times 0.001\,91$
- **(j)** $\dfrac{0.009\,12}{3.121}$

4 This lake is drawn to scale.
Make a rough estimate of its area.

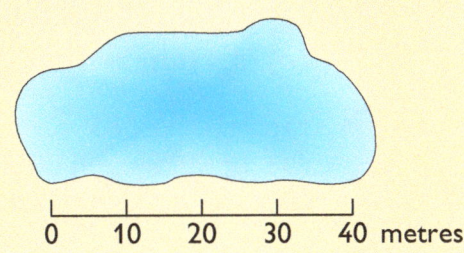

Finishing off

Now that you have finished this chapter you should be able to:

- write numbers to 1 and 2 decimal places
- avoid rounding errors when using your calculator
- express a decimal answer in an appropriate form
- change mixed units to decimals when using your calculator
- make a rough check of a calculation.

Review exercise

1 Choose the smallest from the following sets of numbers.
  (a) 1.3 and 1.29   **(b)** 15.89 and 15.98   **(c)** 0.123 and 0.231
  (d) 4.8, 4.75 and 4.745   **(e)** 0.12, 0.012, 0.021 and 0.21

2 Add 0.01 to each of the numbers below.
  (a) 5   **(b)** 6.4   **(c)** 3.45   **(d)** 1.39   **(e)** 99.99

3 Write the following correct to the number of decimal places in the brackets.
  (a) 34.34 (1)   **(b)** 14.45 (1)   **(c)** 1.646 (1)   **(d)** 1.646 (2)
  (e) 0.137 (2)   **(f)** 0.95 (1)   **(g)** 0.4962 (2)   **(h)** 0.998 (2)

4 For each of the following

  (i) do a rough check   **(ii)** do the calculation on your calculator

  (iii) then give the right answer.
  (a) 29.2×1.98 **(b)** $\dfrac{30.39}{11.6}$ **(c)** 621×0.0399 **(d)** $\dfrac{512 \times 32}{1024}$

5 Write
  (a) 7 litres 20 millilitres in litres only
  (b) 16 hours 35 minutes in hours only
  (c) 12 minutes 16 seconds in minutes only
  (d) 6 days 14 hours in days only
  (e) €5 and 4 cent in € only.

6 Write
  (a) €4.09 in euro and cent   **(b)** 6.35 days in days and hours
  (c) 8.3 hours in hours and minutes   **(d)** 2.9 km in km and metres
  (e) 3.4 minutes in minutes and seconds **(f)** 5.25 days in days and hours.

7 Find the cost of
  (a) 25 litres and 80 millilitres of petrol at 79 cent per litre
  (b) a phone call lasting 15 minutes 56 seconds at 19 cent per minute
  (c) 9 m 12 cm of ribbon at 49 cent per metre.

8 Give an appropriate answer to each of the following:

(a) What is the price of 1 video?

(b) How much washing powder is used for 1 wash?

(c) A bicycle wheel makes 15 revolutions over a distance of 30 metres. How far does it travel in one revolution?

(d) Paulo takes 3 hours and 20 minutes to walk a distance of 16 km. How far does he walk in 1 hour?

9 Find accurate answers to the following:

(a) $(13 \div 7) + (8 \div 7)$

(b) $\sqrt{2} \times \sqrt{8}$

10 Jane has measured the distance around a square.
It is 15.3 cm to 1 decimal place.

(a) Work out the length of one side of the square.
Give your answer to a sensible degree of accuracy.

(b) Use your answer to find the total distance around the square.
Why is this not 15.3 cm?

Investigation   Many fractions cannot be written as exact decimals.

The decimals are called **recurring decimals**.

Example $\frac{1}{3} = 0.333\,33$ *The 3 repeats or recurs.*

$\frac{2}{7} = 0.285\,714\,285\,714$ *The 285 714 all recur.*

Place a dot above the number that recurs.

This can be written as $0.\dot{3}$

This can be written as $0.\dot{2}85\,71\dot{4}$

Place dots above the numbers at the start and end of the recurring pattern.

1 Write the following fractions as decimals correct to

(i) 1 decimal place   **(ii)** 2 decimal places   **(iii)** 3 decimal places

$\frac{2}{3}$ $\frac{3}{7}$ $\frac{5}{11}$ $\frac{4}{13}$ $\frac{2}{9}$

2 Try some more fractions of your own.
Describe any patterns that you discover.

6 Using variables

Simplifying expressions

Tim buys 4 apples and 6 bananas.
At home he has 3 apples
and 1 banana.
Look at what Tim has written
for the total cost of the fruit.

This is called an **expression**.

Accounts
shop   at home
$4a + 6b + 3a + 1b$

Write 4a not a4

You usually write this as b not 1b

3a means 3 × a

Tim **simplifies** the expression
by **collecting like terms**.

$$4a + 6b + 3a + b = 4a + 3a + 6b + b$$
$$= 7a + 7b$$

? What do a and b stand for? Can you simplify this any further?
  What is the total cost of the fruit when
  (i) $a = 12$ $b = 20$ (ii) $a = 20$ $b = 30$?   (where a and b are in cent)

? Simplify $5b - 4b$. How do you write $1b$?
  Simplify $6a - 4a - 2a$. How should you write $0a$?

Tracey finds an expression for the perimeter of the triangle below.
(The measurements are in centimetres.)

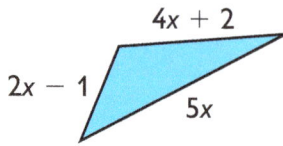

4x + 2
2x − 1
5x

$$\text{Perimeter} = 5x + 2x - 1 + 4x + 2$$
$$= 11x + 1$$

? What is the perimeter of the triangle when $x = 4$?
  Does it make sense for x to be less than $\frac{1}{2}$?

? The perimeter of the triangle is 34 cm. What is the value of x?
  What are the lengths of each side?

Task

1 Find and simplify expressions for the perimeters of the following shapes:

(a)

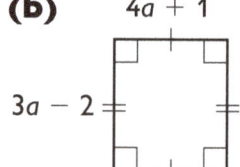

a + b
4a − 2
2b − 3

(b)

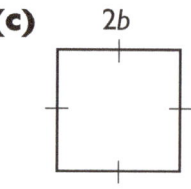

4a + 1
3a − 2

(c)

2b

What is the perimeter of each shape when $a = 4$ and $b = 6$?

2 Look at this cuboid made out of straws.
  (a) Write down and simplify an expression for the total length
  of straw.
  (b) The total length of straw needed is 48 cm.
  What are the possible whole number values of x, y and z?
  Which set of values gives the greatest volume?

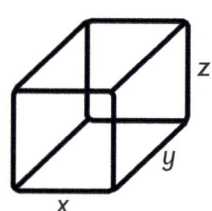

z
y
x

Exercise

1. Simplify the following expressions:
   (a) $a + a + a + a + a$
   (b) $5m + 6m + 12m + 8m$
   (c) $3b + 7 + 2b + 5b - 4$
   (d) $7g - 5 + 3 - 3g - g + 5$
   (e) $5y - 3x - 4y - 2x + 3y$
   (f) $10r - 3p + 6p - 7r + 8p - 3r$
   (g) $3x + 7x - 4 - 6x + 1$
   (h) $5r - 6t + 7t - 2r - 3r$

2. (a) (i) Write down an expression for the sum
          of the angles in this triangle.
      (ii) Simplify your expression.
   (b) What do the angles in a triangle add up to?
   (c) What is the value of x?
   (d) Write down the value of each of the three angles.

   $4x$ $2x$ $3x$

3. Jenny buys 3 apples and 6 bananas.
   (a) Write down an expression for the cost of the fruit.
       Use a for the cost in cent of an apple and b for the cost in cent of
       a banana.

   Bananas cost 15 cent each.
   (b) Write down an expression for the cost of the fruit.

   Jenny pays €1.20 for the fruit.
   (c) How much is an apple?

4. Look at these algebra walls.

   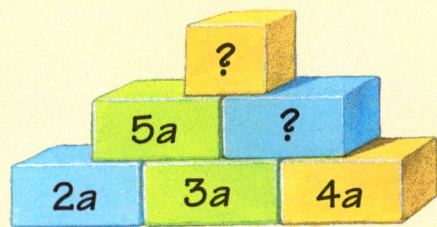

   The expression in each brick is the sum of the two bricks underneath it.
   Copy each algebra wall and find the missing expressions.

5. Dan and Rachel have simplified the expression

   $5a - 7b - 4b + 7a - a.$

   The answer is $11a - 11b$

   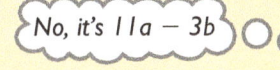
   No, it's $11a - 3b$

   (a) Who is right? Show all your working.
   (b) How could you check that you have the correct answer?

More expressions

Alison wants to find expressions for the areas of these shapes:

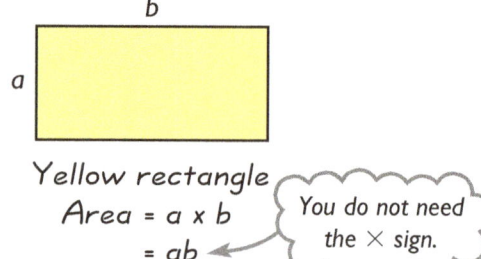

She writes: *Green square*
            Area = r x r

*r × r is written as r². You say this as **r squared**.*

            = r²

Yellow rectangle
Area = a x b

You do not need the × sign.

            = ab

? **The area of the square is 36 cm².**
What is the value of *r*?

? **Is *ab* the same as *ba*?**
Check your answer by substituting *a* = 4 and *b* = 5 into both expressions.
Is 3*ab* the same as *b*3*a*? How can you tell?

❝ Do the right thing!

Step 1 You should always write letters
           in alphabetical order.
           So write *ab* not *ba*.

Step 2 You should write numbers first
           then letters.
           So write 4*ab* not *ab*4 or *a*4*b*.

Alison writes down the volume for this blue cube:

            Volume of blue cube = $p \times p \times p$

*p × p × p is written as p³. You say this as **p cubed**.*

            = p^3

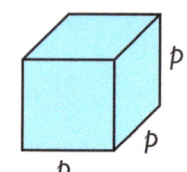

$n \times n \times n \times n$ is written as n^4. You say this as ***n* to the power of 4**.

? **What is the volume of the cube when *p* = 3?**
How would you say $n \times n \times n \times n \times n$?

Task

1 Find the value of these
      expressions when $n = 2$.
      (a) n^2 **(b)** n^3
      (c) n^4 **(d)** n^5

2 Use your answers to work
      out the value of
      (a) $n^4 \times n$ **(b)** $n^2 \times n^3$
      (c) $n \times n^3$ **(d)** $n^2 \times n^2$
      (e) $n^5 \div n$ **(f)** $n^5 \div n^2$
      (g) $n^5 \div n^3$ **(h)** $n^4 \div n^2$

3 Which of the expressions in Question 2
      match the expressions in Question 1?

4 Look at the cuboid.
      Find the volume when
      (a) $x = 3$ $y = 6$
      (b) $x = 2$ $y = 9$
      (c) $x = 1.5$ $y = 20$

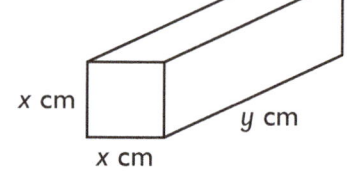

? **How would you simplify (i) $6n^2 - 2n^2$ (ii) $4n^2 + 2n^2$ (iii) $2n^2 + 3n^2 - n^2 - 4n^2$**
How can you check you are right?

Exercise

1 How should you write the following expressions?

(a) $p \times r$ **(b)** $3 \times f$ **(c)** $g \times 3 \times h$ **(d)** $n \times n$ **(e)** $g \times 5$

(f) $p \times 6$ **(g)** $a \times 2 \times c$ **(h)** $2 \times m \times m$ **(i)** $x \times 3 \times x$ **(j)** $d \times 3 \times e$

2 Simplify the following expressions.

(a) $n^2 + n^2$ **(b)** $x^2 + x^2 + x^2$ **(c)** $p^2 + p^2 + p^2 + p^2$

(d) $4q^2 + 2q^2$ **(e)** $3f^2 + 2f^2 + f^2$ **(f)** $5c^2 - 2c^2$

(g) $8m^2 - 3m^2$ **(h)** $7t^2 + 2t^2 - 3t^2$ **(i)** $3r^2 + 2r^2 - r^2$

3 **(a)** What are the values of these expressions when $a = 3$?

     (i) a^2 **(ii)** $a^2 + a^2$ **(iii)** $4a$

     (iv) $a + 2a + a + 2a$ **(v)** $2a^2$ **(vi)** $6a$

   (b) Simplify $a^2 + a^2$

   (c) Give meanings for (i)–(vi) in this diagram: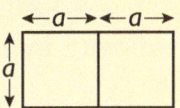

4 This is a magic square because the sum of the expressions in each row, column and diagonal is the same.

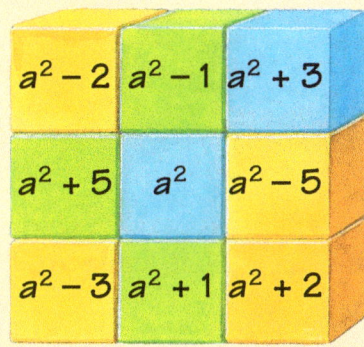

   (a) Find the sum of each row, column and diagonal
when **(i)** $a = 2$ **(ii)** $a = 5$.

   (b) Check that the square is still magic for a third value of a.

   (c) Find in terms of a the sum of each row, column and diagonal.

   (d) Is the square magic for any value of a?

5 Kelly is investigating powers.
She writes

$$3^3 \times 3^2 = 3 \times 3 \times 3 \times 3 \times 3 = 3^5$$

So $\quad p^3 \times p^2 = p \times p \times p \times p \times p = p^5$

Simplify the following.
Set out your work like Kelly did.

(a) (i) $3^2 \times 3$ **(b) (i)** $5^2 \times 5^2$ **(c) (i)** $4^3 \times 4$

       (ii) $p^2 \times p$ **(ii)** $p^2 \times p^2$ **(ii)** $p^3 \times p$

Brackets and factorising

Using brackets

Phil and Debbie are playing a game of *Algebra Snap*.
They both have a set of cards with expressions on them.

They take turns in laying down cards.
They can say 'Snap!' when both their cards show matching expressions.

The person who calls 'Snap!' first collects the cards that have already been laid.
The winner is the person with the most cards.

Here is Debbie's first card: **6(2a + 3)**   *This means 6 × (2a + 3).*

Here is Phil's first card: **18 + 12a**

Debbie **expands the bracket**. She multiplies every term inside the bracket by the number outside the bracket.
She writes:

$$6(2a + 3) = 6 \times 2a + 6 \times 3$$
$$= 12a + 18$$

 Is this the same expression as on Phil's card?

Another card is: **3(4a + 6)**   **Does this match the expression on Phil's card? Write down another expression using brackets which matches Phil's card.**

Factorising

One of Phil's cards is: **36a − 24**

Phil **factorises** this expression. He writes it using brackets.

 What is the largest number that is a factor of both 36 and 24?

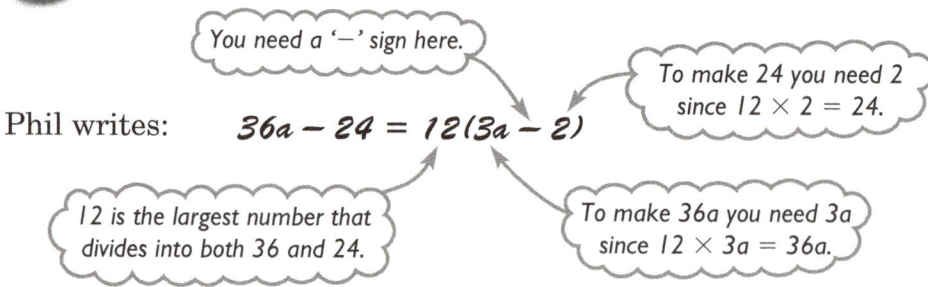

Phil writes: $36a - 24 = 12(3a - 2)$

You need a '−' sign here.

To make 24 you need 2 since 12 × 2 = 24.

12 is the largest number that divides into both 36 and 24.

To make 36a you need 3a since 12 × 3a = 36a.

 You can write 36a − 24 as 2(18a − 12). This is partly factorised. What other ways can you partly factorise 36a − 24?

We say that Phil has **factorised** 36a − 24 **fully**.
Phil has chosen 12 to go outside the brackets.
This is the largest number which is a factor of both 36 and 24.

 **Task**

Make a set of your own *Algebra Snap* cards.
Play your game of *Algebra Snap* with a partner.

Exercise

1 Sam is using brackets to help her with multiplication.

Here is a copy of her work.

Work out the following using the same method as Sam.

$3 \times 27 = 3 \times (20 + 7)$
$= 3 \times 20 + 3 \times 7$
$= 60 + 21$
$= 81$

(a) 5×23 **(b)** 6×32

(c) 7×47 **(d)** 8×84

2 Expand the following brackets:

(a) $5(3x + 7)$ **(b)** $4(2a + 5)$ **(c)** $8(3 + 2c)$

(d) $3(5t - 1)$ **(e)** $7(1 - 2n)$ **(f)** $6(3 - 2y)$

(g) $100(a + 2)$ **(h)** $50(2a + 4)$ **(i)** $25(4a + 8)$

3 Factorise each of the following expressions *fully*.
Write down any ways of factorising them *partly* as well.

(a) $3x + 6$ **(b)** $2x + 10$ **(c)** $10x + 2$

(d) $10x - 2$ **(e)** $5x + 15$ **(f)** $15f - 45$

(g) $30s - 40$ **(h)** $12b + 6a$ **(i)** $48 - 24a$

4 Factorise the following expressions fully.

(a) $4a + 8$ **(b)** $12g + 4$ **(c)** $18k - 12$

(d) $24f - 18g$ **(e)** $15r + 3$ **(f)** $21t - 7$

(g) $4h + 16$ **(h)** $8i + 4$ **(i)** $8j - 8$

5 John has $3n - 1$ sweets and Mary has $3(n - 1)$ sweets.
Who has more sweets? Explain your answer clearly.

6 Simplify the following expressions:

(a) $5(b - 2) + 4(2 - b)$ **(b)** $3(d + 2) - 2(d + 1)$

7 Match up the following *Algebra Snap* cards into groups.
Which card is the odd one out?

$15a + 5$

$6(2a - 1)$

$12a - 6$

$18 - 12a$

$3(5 - 2a)$

$10a - 2b - 6 + 2(a + b)$

$6(3 - 2a)$

$6(3a - 2) + 3(2 - 2a)$

$8a + 10 + 7a - 5$

$5(a + 1) + 10a$

$3(4a - 2)$

$5(1 + 3a)$

Finishing off

Now that you have finished this chapter you should be able to:

- know the meaning of the terms expression, simplify, factorise and expand
- simplify an expression
- know that $6a$ means $6 \times a$

  a^2 means $a \times a$

  a^3 means $a \times a \times a$
- factorise an expression
- expand a bracket.

Review exercise

1 Simplify the following expressions:

(a) $4x + 5y + 3x + 7y$ (b) $3k + 5 + 2k + 4$

(c) $5a - 2a + 3b - b$ (d) $5f + 6 - 5f - 7$

(e) $5s + 3s + 5t - 4s + 6t$ (f) $4g - 6h - 7g + 7h + 2g$

(g) $2x + 4 + 5x - 4$ (h) $3x + y - 2x - y$

(i) $5a - 4b + 6a + 5b + 9a - b$ (j) $5p - 3 + 2p + 4 - 3p + 6$

2 How should you write the following expressions?

(a) $5 \times 4 \times f$ (b) $s \times s$ (c) $5 \times m \times m$

(d) $3 \times d \times e$ (e) $b \times 4 \times a$ (f) $g \times 6 \times g$

(g) $2 \times 3 \times x$ (h) $5 \times y \times 7$ (i) $p \times p \times p$

3 Expand the following brackets:

(a) $5(2 + c)$ (b) $3(x - 6)$ (c) $7(p + r)$

(d) $4(2d + 3)$ (e) $4(b + 8)$ (f) $7(2r - 1)$

(g) $3(1 - k)$ (h) $5(2a - 3b)$ (i) $5(x + y)$

(j) $3(2x - y)$ (k) $6(3a - 4b)$ (l) $12(3q - 5r)$

4 Factorise each of these expressions *fully*.
Write down any ways of factorising them *partly* as well.

(a) $18a + 6$ (b) $12d - 4$ (c) $24b - 36c$

(d) $88h - 44$ (e) $15k - 12$ (f) $10 - 30f$

5 **(a)** Write down and simplify an expression for the sum of the angles in this quadrilateral.

(b) (i) What do the angles in a quadrilateral add up to?

(ii) What is the value of p?

(c) Use your value of p to find the size of each of the four angles.

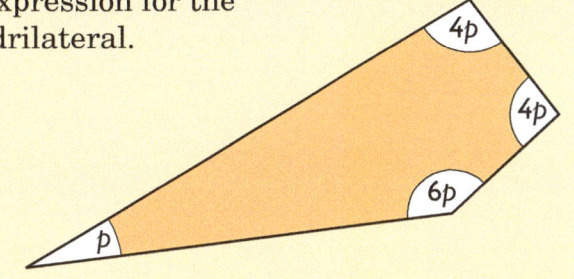

6 **(a)** Find and simplify an expression for the perimeter of this kite.

(b) What is the perimeter when $n = 3$?

(c) The perimeter of the kite is 40 cm.
(i) What is the value of n?
(ii) What is the length of the shortest side of the kite?

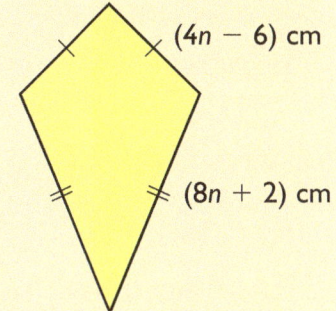

$(4n - 6)$ cm

$(8n + 2)$ cm

7 Fiona is having a party. She buys 5 bottles of coke and 7 bottles of lemonade. Bottles of lemonade cost 60 cent each.

(a) Write down an expression for the cost of the drinks in cent.
Use c for the cost of a bottle of coke.

(b) Fiona spends €6.70 on the drinks. How much is a bottle of coke?

Investigation

— Family Fortune —
The TV Quiz Game

- Family teams of 5, with at least 2 children
- 1 question to each team member; choose 'easy' or 'hard'
- Each team is given €1000 'playing money'
- For each correct answer: Easy = €400
  Hard = €1000
- For each wrong answer: deduct €200

Easy questions right, e

Hard questions right, h	0	1	2	3	4	5
0						3000
1			2400			
2						
3						
4						
5	6000					

Winnings Table (€)

1 Copy and complete the Winnings Table.

The McTaggart family get e easy questions right and h hard questions right.

2 Write down an expression for the number of questions that the McTaggart's get wrong.

3 Find an expression for the McTaggart's winnings.

4 Check that your expression gives the numbers in your Winnings Table.

7 Construction

Do you remember?

Constructing triangles

Look at the two triangles opposite.

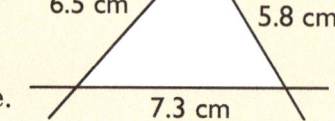
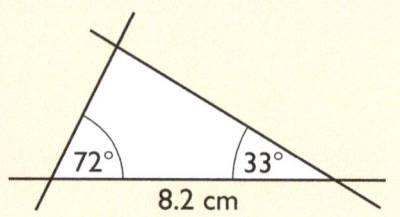

1 Make an accurate construction of each triangle.

2 Write down the steps you followed to draw them.

Scale drawings

THE AVONFORD STAR

Dangerous equipment in playground

Following several accidents the popular 'Pole Swing' has been removed from the play area in the civic park.

'Kids were not using it sensibly,' said Leisure Manager John Stubbs. Owain Wilson, aged 8, said: 'My friend climbed to the top of the pole, trapped his fingers then fell to the ground.'

Engineer's sketch

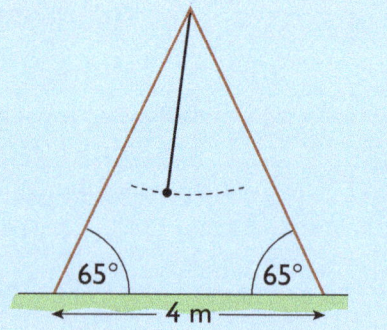

 Look at the diagram. Estimate the height of the triangular frame.

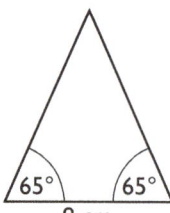 **Task**

Drawing to scale

1 Construct a triangle with a base of 8 cm and two angles of 65°.
Your triangle is a **scale drawing** of the frame of the swing.

The **scale** of the drawing is 2 cm to 1 m.

2 cm on the scale drawing represents 1 m on the frame.

2 Using your scale drawing what is the height of the frame?

 How close was your estimate?

3 What is the total length of metal tubing used to make the frame?

The pole is 3.5 m long.

4 Use a pair of compasses to draw the path of the bottom of the pole as it swings.
Show the pole in two different positions of its swing.

 How do you use a ratio to describe the scale of your drawing?

Exercise

1 **(a)** Construct triangles with sides of the following lengths.
   (i) 4.5 cm, 8 cm, 9 cm   **(ii)** 2.5 cm, 6 cm, 6.5 cm
   (iii) 4 cm, 7.5 cm, 8.5 cm   **(iv)** 3 cm, 3 cm, 4 cm

(b) Which of the triangles are right-angled?

2 **(a)** Make accurate constructions of these sketched triangles.

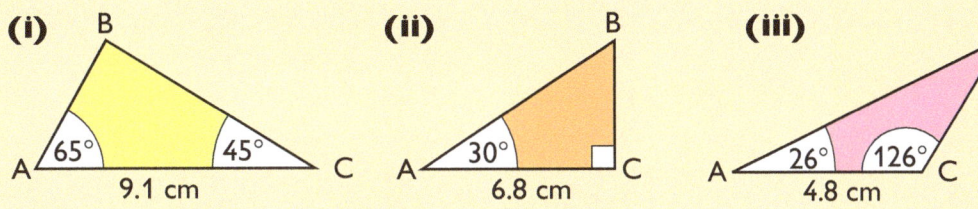

(i) B, 65°, 45°, A, 9.1 cm, C   **(ii)** B, 30°, A, 6.8 cm, C   **(iii)** B, 26°, 126°, A, 4.8 cm, C

(b) For each triangle measure the lengths of AB and BC.

(c) For each triangle measure the angle at B.
   Check the accuracy of your construction by *calculating* angle B.

3 Look at the plan of a rectangular field.
It is *not* to scale.

A diagonal path goes across the
field from A to C.

Construct a scale diagram to find
the length of the path.

Choose your own scale.

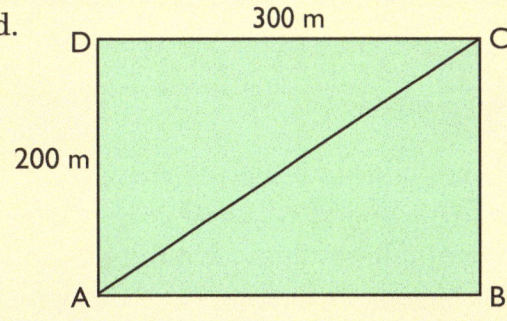

D — 300 m — C
200 m
A — B

4 Make accurate constructions of these triangles.

(a)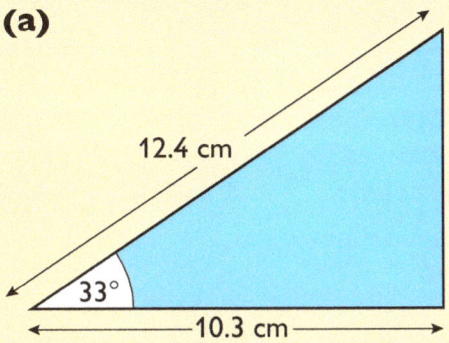

12.4 cm
33°
10.3 cm

(b)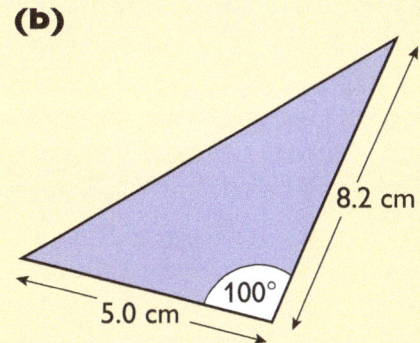

8.2 cm
100°
5.0 cm

Measure and label the length of the unlabelled side for each triangle.

More scale drawings

Angles of elevation and depression

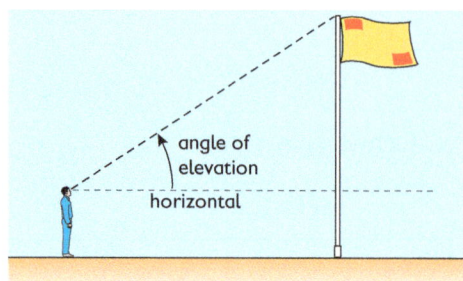

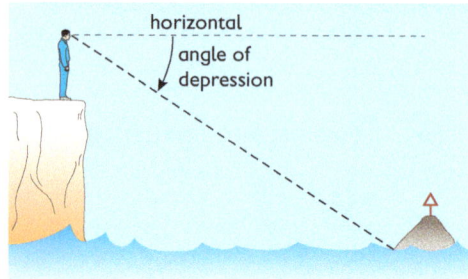

An **angle of elevation** is measured **upwards** from the horizontal.

 You want to find the height of the flagpole without climbing it. You decide to do a scale drawing. What information do you need?

An **angle of depression** is measured **downwards** from the horizontal.

 The angle of depression of the buoy from the man's head is 40°. What is the angle of elevation of the man's head from the buoy?

Scale drawings of compass bearings

Task

This diagram is a scale drawing of a helicopter journey.

The helicopter flies from Carlisle to Birmingham, then Birmingham to Norwich.

1 How far is it from Carlisle to Birmingham?
2 How far is it from Birmingham to Norwich?
3 What is angle *X*?
4 Make an accurate copy of the scale drawing.
5 The helicopter flies back to Carlisle *from* Norwich.
   Show this return journey on your diagram.
6 What is the distance of Carlisle from Norwich?
7 What is the bearing of Carlisle from Norwich?

The next day the helicopter flies East from Carlisle to Newcastle. The distance from Carlisle to Newcastle is 90 km.

8 Show the journey on your scale drawing.
9 How far is Newcastle from Norwich?

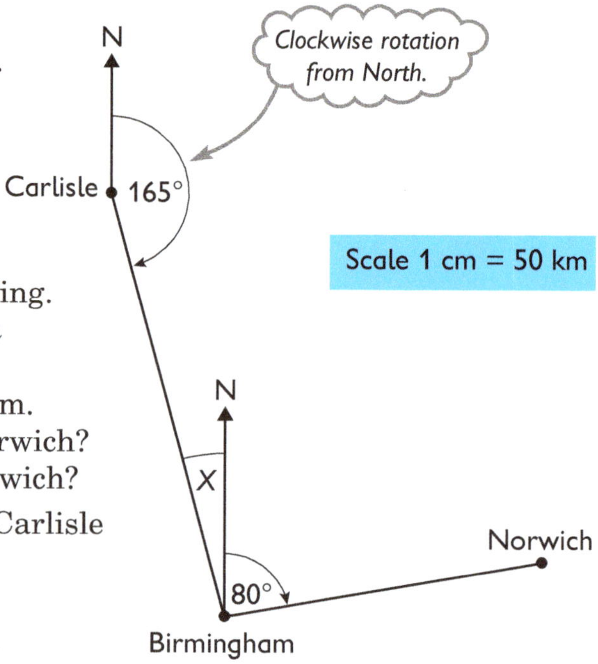

N

Clockwise rotation from North.

Carlisle • 165°

Scale 1 cm = 50 km

N

X

Norwich

80°

Birmingham

You can use scale drawings to solve problems involving angles, distances and bearings.

Exercise

1 Dave is a rock climber.
He needs to know the vertical height
of the cliff face.
He marks a point A on the ground 50 m
from the bottom of the cliff.

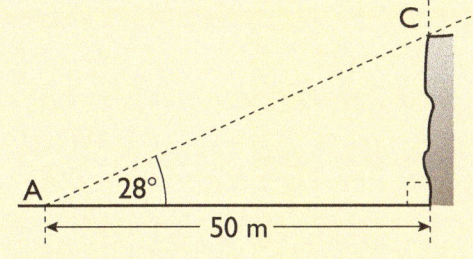

The angle of elevation from A to the
top of the cliff C is 28°.

    (a) Using a scale of 2 cm to 10 m, construct the right-angled triangle as a
    scale drawing.

    (b) Use your scale drawing to find the height of the cliff.

2 An aeroplane flies from Barnley
to Shippon, then from Shippon
to Hutford.

    (a) Make an accurate drawing of
    the journey using a scale
    of 1 cm to 10 km.

    (b) Use your diagram to find the
    bearing and distance the
    aeroplane flies to return directly to Barnley from Hutford.

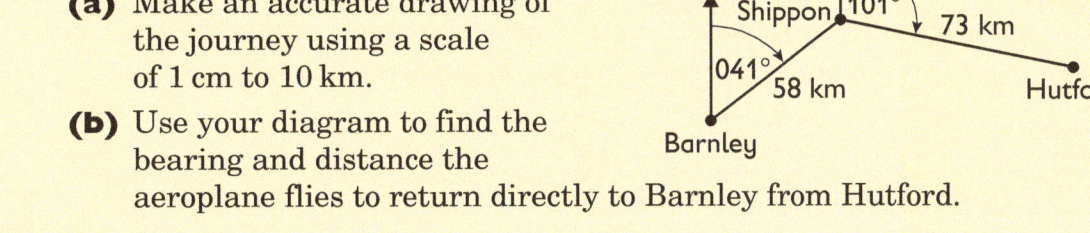

3 A lighthouse is 120 m tall.
The angle of depression of a
tanker from the top of the
lighthouse is 9°.

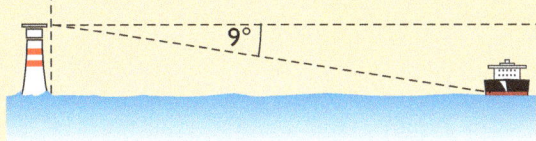

    (a) Using a scale of 2 cm to 100 m,
    construct a scale drawing.

    (b) Use your scale drawing to find the distance between the tanker and
    the base of the lighthouse.

4 A coastguard sees a yacht in distress.

    The yacht is 5 km from the coastguard on a bearing of 283°.

    A rescue helicopter is 6 km from the
    coastguard station on a bearing of 064°.

    (a) Make an accurate scale drawing
    showing the positions of the yacht,
    helicopter and coastguard station.
    Use a scale of 1 cm to 1 km.

    (b) How far is the helicopter from
    the yacht?

    (c) The helicopter flies straight to
    the yacht. Find the bearing on which
    the helicopter flies.

    (d) The helicopter flies at 150 km/h.
    How long does it take to reach the yacht?

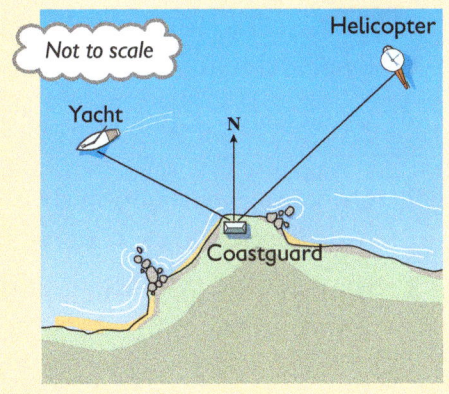

Perpendiculars

THE AVONFORD STAR

New footpath for walkers

A new footpath has opened across Avonford Meadows.
It starts at the lightning tree in the hedge.
From here it goes at right angles to the hedge
until it reaches the other side of the meadow.

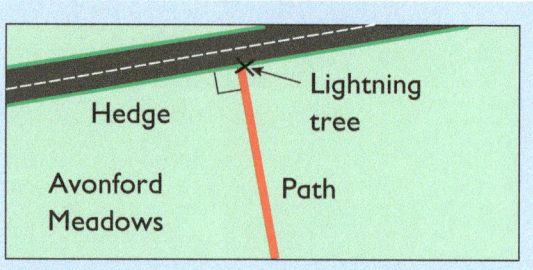

A line at right angles to another is called a **perpendicular**.

You can draw the map *without* using a protractor.

" Do the right thing!
Construct a perpendicular from a point on a line

You will need a pencil, a pair of compasses and a straight edge.

STEP 1 Mark T on a line.

STEP 2 Mark A and B.
Each point is
the same
distance from T.

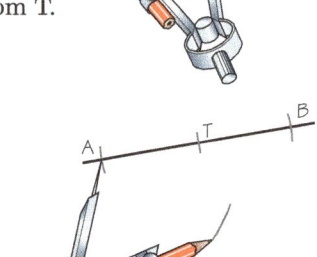

STEP 3 Open the
compasses
wider.
With the
point on A
draw an arc.

STEP 4 *Do not adjust
the compasses.*
With the point
on B draw an
intersecting arc.

STEP 5 Draw a line
through T and
the intersection
X.

> Line TX is a perpendicular to AB from point T.

Task

1 Check the angle at T with a protractor.

? **What can you say about lengths AX and BX?**

2 On your construction draw another pair of
intersecting arcs, centred at A and B.

3 Repeat this for a third pair of arcs.
What do you notice?

*Make sure that the compasses
are adjusted to the same
radius for each arc.*

TX is the **locus** of points **equidistant** from A and B.
A locus is a set of points which obey a rule.

*Equidistant means the
same distance.*

Exercise

1 **(a)** Draw a line.
Mark a point T on it.

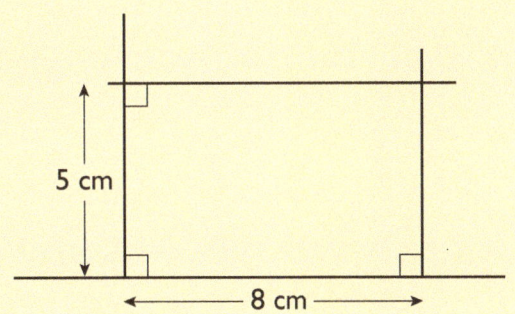

(b) Use a straight edge and a pair of
compasses to construct a line
from T which is perpendicular
to the line.

A straight edge

A ruler
What is the difference?

(c) Use a protractor to check that you
have constructed a right angle.

*Do not use a protractor
for Questions 2 and 3.*

2 Use a ruler and a pair of compasses
to make an accurate construction
of this rectangle.

Measure the length of a diagonal.

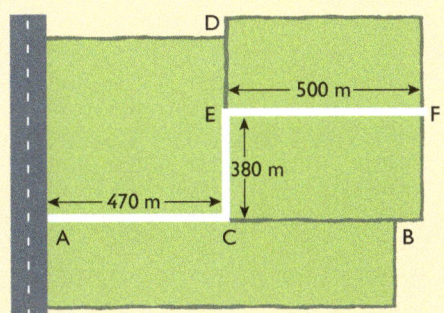

5 cm

8 cm

3 Look at this diagram.
It shows a path ACEF following the
edges of rectangular fields.

Construct the *path* to scale.
Use a scale of 1 cm to 100 m.

Use your scale drawing to find the
straight line distance from A to F.

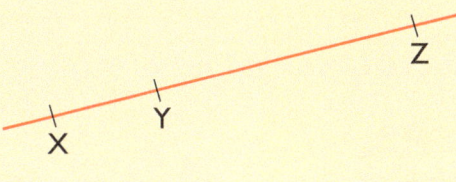

D

500 m

E F

380 m

470 m

A C B

4 Draw a straight line.
Mark and label three points
X, Y and Z on the line.
Construct perpendiculars from
each point.
What can you say about your three perpendiculars?

Z

Y

X

Activity **1** Use LOGO to draw the path you constructed in Question 3.

2 Use LOGO to construct a path with 90° turns.
The path starts and finishes at the same place.

? **What is the minimum number of legs your path can have?**
Is there a maximum number of legs?

More perpendiculars

Look at the diagram. The rabbit wants to reach the hedge in the shortest distance.

R is the position of a rabbit in a field. The rabbit needs to hide quickly.

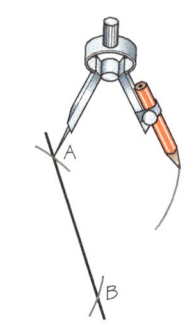

Hedge

? **What path should the rabbit take?**
How do you construct this path on the diagram?

" Do the right thing!
Construct a perpendicular from a point to a line

You will need a pencil, a pair of compasses and a ruler.

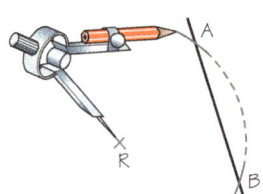

STEP 1 Draw a line (the hedge). Mark R away from the line.

STEP 2 Put the point of the compasses on R. Draw an arc to intersect at A and B.

STEP 3 With the point on A draw an arc on the opposite side of the line from R.

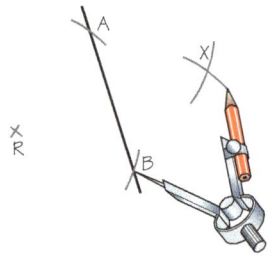

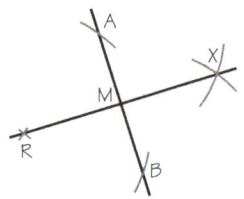

Line RM is a perpendicular from point R to line AB.

STEP 4 *Do not adjust the compasses.* With the point on B, draw another arc to intersect at X.

STEP 5 Draw a line through R and X. Label M, the intersection with A and B.

? **Measure lengths AM and BM. Explain what you notice.**

Task

You are going to construct the perpendicular bisector of a line.

Bisect means to cut in half.

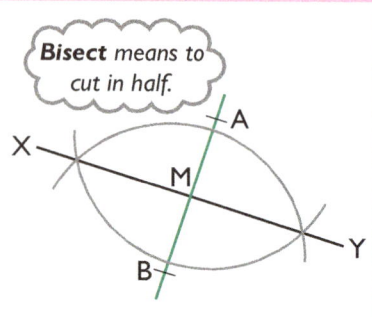

1 Copy the green line with A and B marked.
2 Write down the steps to construct XY and follow them.
3 Check that lengths AM and BM are equal.
4 What is the angle at M?

XY is the perpendicular bisector of line segment AB

? **Describe a rule for the locus XY.**

Exercise

1 **(a)** Draw a straight line PQ of length 9 cm.
   (b) Construct the perpendicular
      bisector of PQ.
      Point M is the midpoint of PQ.
   (c) Use a protractor to check that the
      angle at M is a right angle.
      Check that PM = QM = 4.5 cm.

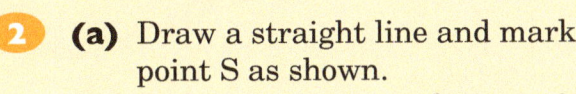

2 **(a)** Draw a straight line and mark
      point S as shown.
   (b) Follow the steps on the opposite
      page to construct a perpendicular
      from S to the line.
      Check the angle with a protractor.

×S

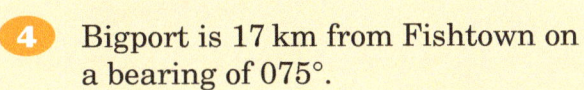

3 A and B are the positions of two radio beacons.
   An aeroplane flies so that it is
   always equidistant from each beacon.
   (a) Draw two crosses to represent
      the beacons.
   (b) Construct a line to represent
      the aeroplane's path.
   (c) What is the locus of the path?

A×

B×

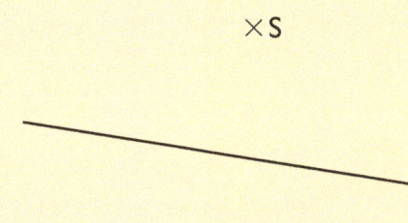

4 Bigport is 17 km from Fishtown on
   a bearing of 075°.
   Mullford is 48 km from Fishtown on
   a bearing of 023°.
   (a) Construct a scale diagram showing
      the positions of the three towns.
      Choose your own scale.
   A tanker sails past in the direction
   of the arrow.
   In order not to run aground, its path must always be equidistant from
   Bigport and Mullford.
   (b) Use a pair of compasses and ruler to construct the line of the tanker's
      path on your diagram.
   (c) Find the bearing of the tanker's course.

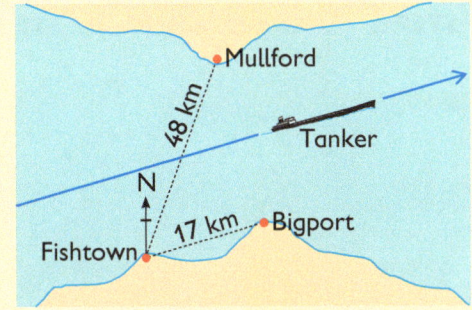

Activity **1** Use a pair of compasses and ruler
         to construct a pair of parallel lines
         as shown.

      2 Use a pair of compasses and a ruler to
         construct another line, equidistant
         between the first two lines.

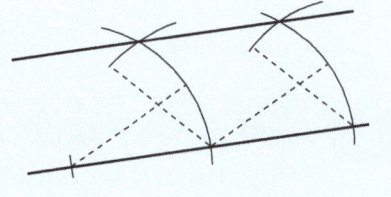

Bisecting an angle

Look at this plan of a new street.

The extra bit of land between numbers 91 and 93 is to be shared equally between the two houses.

Where should the dividing wall go between the two plots of land?
Describe the locus of the wall.

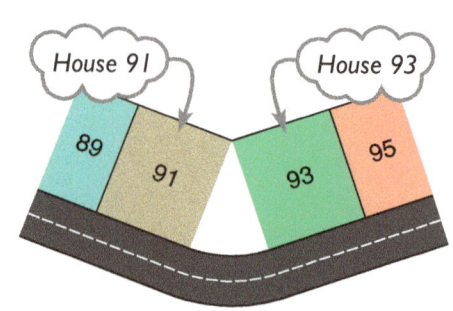

To accurately **bisect** the angle between the two plots you use a pair of compasses.

❝ Do the right thing!
Construct a line to bisect an angle

You will need a pencil, a pair of compasses and a ruler.

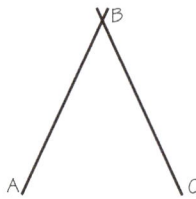

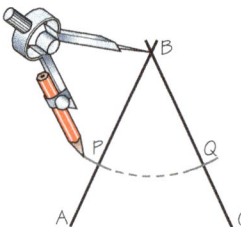

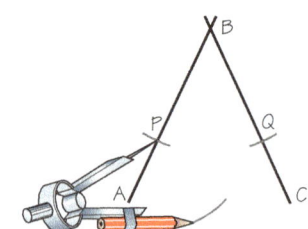

STEP 1 Draw and label an angle ABC.

STEP 2 Mark points P and Q. They are the same distance from B.

STEP 3 With the point of the compasses on P, draw an arc between A and C.

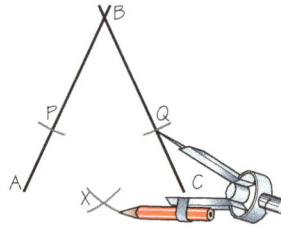

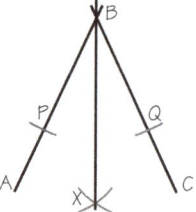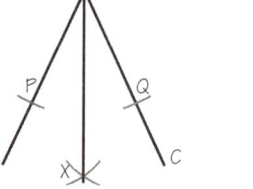

STEP 4 *Do not adjust the compasses.* With the point on Q, draw another arc to intersect at X.

STEP 5 Draw a line through B and X. Use a protractor to check your construction.

Line BX is the bisector of angle ABC

 How does 'ABC' describe the angle in STEP 1?

 Draw another pair of intersecting arcs centred on P and Q. What do you notice?
Why must X be equidistant from P and Q?
Describe a rule for the locus BX.

Exercise

1 **(a)** Use a protractor to draw and label the angles below.

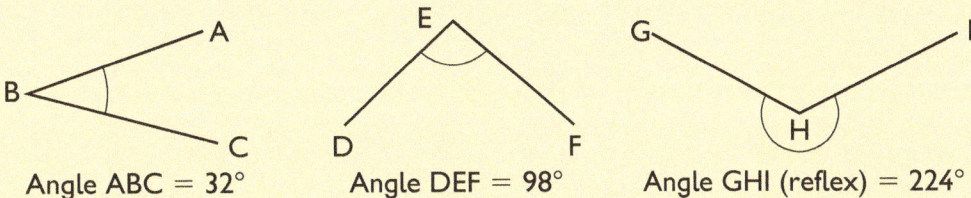

Angle ABC = 32°   Angle DEF = 98°   Angle GHI (reflex) = 224°

(b) Follow the steps on the opposite page to bisect each of the angles described.

(c) Check the accuracy of your constructions with a protractor.

2 **(a)** Make an accurate construction of this isosceles triangle.

(b) Use a pair of compasses and a ruler to construct the line of symmetry of the triangle.

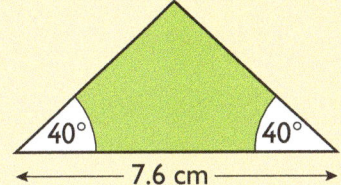

3 **(a)** Make an accurate construction of triangle ABC.

(b) Construct the bisector of angle ABC, so that it intersects side AC. Label the point of intersection X.

(c) Measure lengths AX and BX. Calculate length CX.

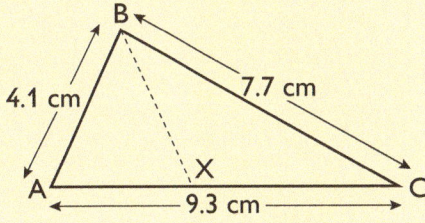

Investigation

1 Draw any triangle.

2 Construct the bisectors of each interior angle of your triangle. What do you notice about the three bisectors?

3 Repeat Parts 1 and 2 with a different triangle.

4 Compare your diagrams with other members of your class.

What can you say about the bisectors of the interior angles of a triangle?

Activity

Construct a compass rose

Before the 20th century navigators used a compass rose to describe direction.

Use a pair of compasses and a ruler to construct one like this.

Use a whole page.

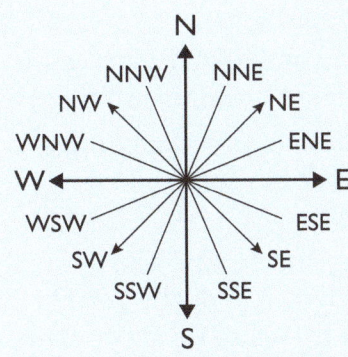

Finishing off

Now that you have finished this chapter you should be able to:

- make accurate constructions of triangles
- construct scale drawings and use them to solve problems
- construct a perpendicular from a point *to* a line and from a point *on* a line
- construct the perpendicular bisector of a line and label the midpoint
- bisect an angle
- describe a locus.

Review exercise

1 **(a)** Make accurate constructions of these triangles.

(i) 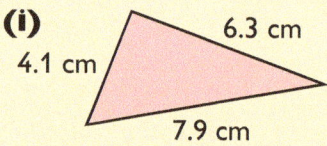   **(ii)** 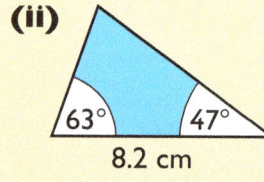   **(iii)**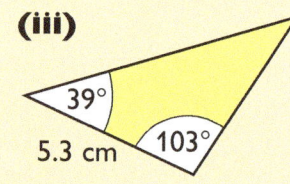

    (iv) a triangle with sides 4.5 cm, 5.8 cm and 11.1 cm.

(b) In each case measure the unknown sides and angles.

2 Habib wishes to find the height of a mast.
At a point 65 m from the base of the
mast the angle of elevation of the top is 39°.

Draw a scale diagram to find the
height of the mast.

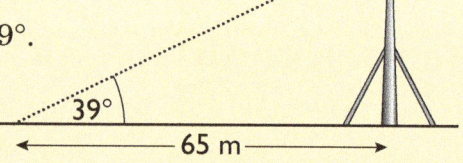

3 Use a pair of compasses and a ruler to construct a square with sides of 7 cm.
Measure the length of the diagonals.

4 Make an accurate construction of
this parallelogram.
Measure the length of each diagonal.

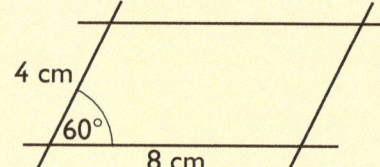

5 A desert patrol leaves base and drives
to an oasis, then from the oasis to a cave.

(a) Make an accurate scale drawing of
the journey using a scale of
1 cm to 10 km.

(b) Use your diagram to find the
bearing and distance of the return
journey from the cave direct to base.

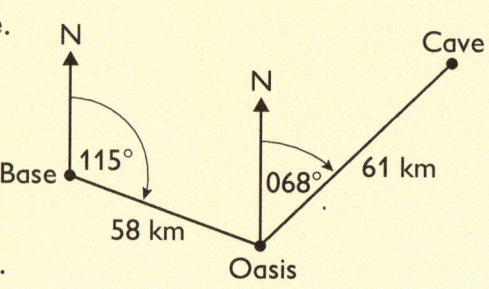

6 Draw two intersecting lines.
Use a pair of compasses and ruler to bisect all four angles.
What do you notice about your bisecting lines?

7 Gordon is walking on the moors.

To check his position he takes compass bearings of two masts.

From a map he knows that the TV mast is 13.5 km North of the Youth Hostel.

The satellite dish is 4.1 km West of the Youth Hostel.

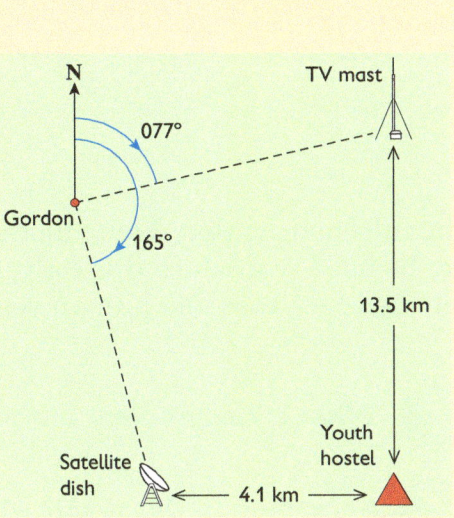

(a) In what direction must he walk to reach the Hostel?

(b) Gordon reckons he can walk at 4 km per hour.
It is 2 pm when he checks his position.
Will he reach the Hostel in time for dinner at 6 pm?

Investigation

1 Draw any triangle.
2 Construct the perpendicular bisector of each side of your triangle.
3 What do you notice about the three perpendiculars?
4 Repeat Parts 1 and 2 with a different triangle.

Compare your diagrams with other members of your class.

5 What can you say about the perpendicular bisectors of the sides of a triangle?

Activity

Look at the diagram.
Rectangle ABCD is a rectangular field.

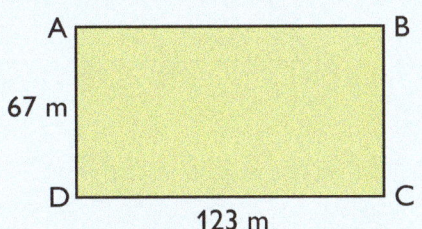

1 Using a scale of 1 cm to 10 m construct a scale drawing of the field.

A treasure chest is buried in the field.
The chest is equidistant from walls AB and AD, and from corners B and C.

2 Use your construction skills to find the location of the treasure.
Mark the place on your diagram and label it with a T.

3 How far is the treasure from corner B?

The information that Jane's mother has given can be used to produce this conversion graph between cat years and human years.

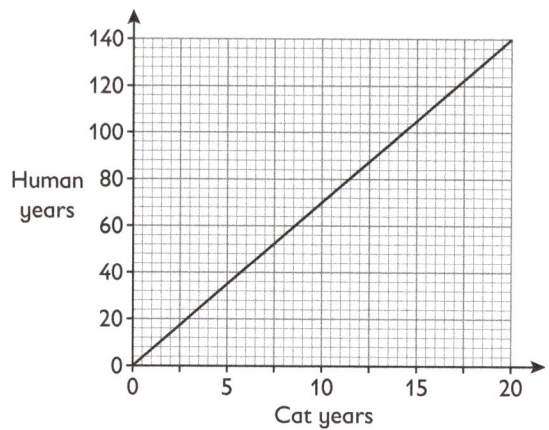

? **What is the gradient of this line?**

? **According to this graph, what is Tabby's human age?**

Jane does not believe her mother. She gets this graph from the vet.

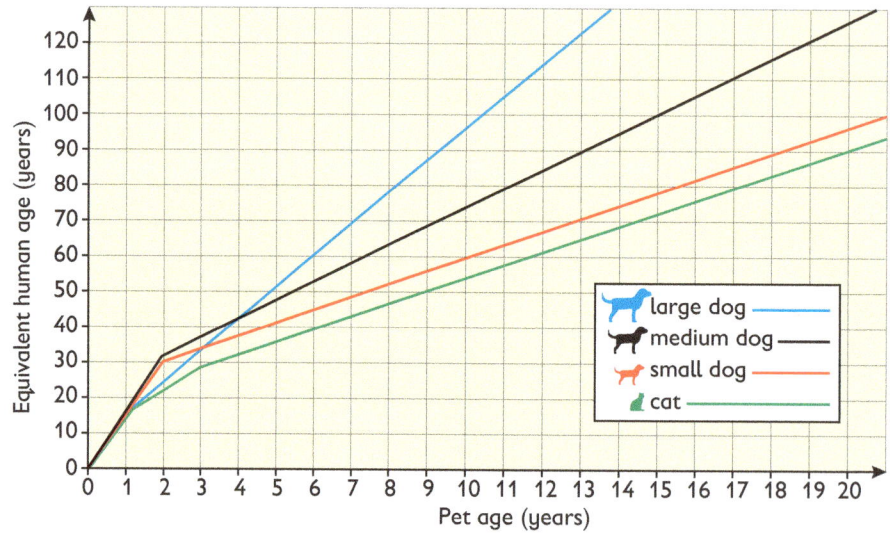

Task

1 How old is Tabby according to the vet's graph?

2 Which is older in human years, a 9-year-old large dog or a 19-year-old cat?

3 Is it ever true that a dog or cat's human age is 7 times its real age?

? **What information do the gradients of the lines on this graph tell you?**

Exercise

1 This graph shows the temperature throughout one day, starting at midnight.

(a) What is the highest temperature?

(b) When is it hottest?

(c) What is the lowest temperature?

(d) When is it coldest?

(e) Between what times is the temperature rising fastest?

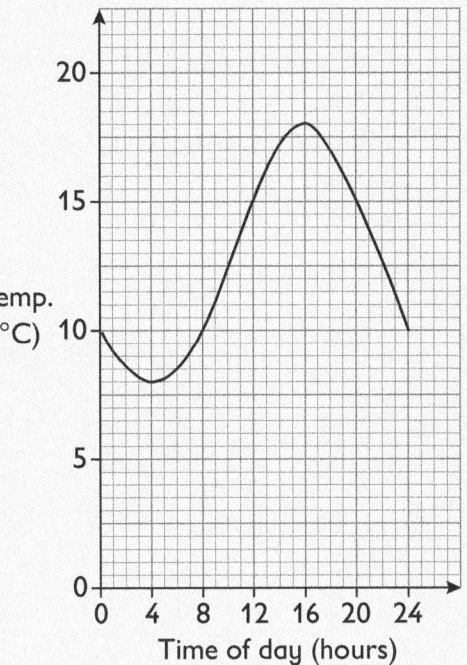

Temp. (°C)

Time of day (hours)

2 These figures give the height of a tree at various ages.

Age (years)	0	5	10	15	20	30	40	50	60	70	75	75
Height (metres)	0	3	10	25	34	40	40	40	40	40	40	0

(a) Draw a graph of the height of the tree against its age.

(b) When is the tree fully grown?

(c) What is the tree's greatest height?

(d) How long does it live?

(e) Between what ages is it growing fastest?

3 This graph shows the volume of water in a bath. For each part of this story, state the times.

(a) The taps are turned on. The bath fills.

(b) Sergio is lying in the bath.

(c) Sergio lets out some of the water.

(d) He puts in more hot water.

(e) He has a final soak.

(f) He gets out and lets out all the water.

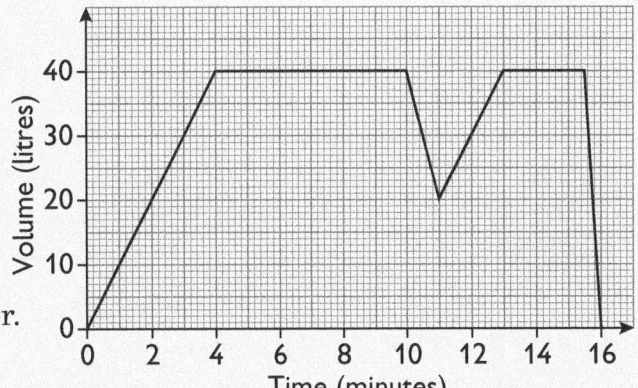

Volume (litres)

Time (minutes)

Travel graphs

Look at this graph showing the height of a
hot air balloon.
It is called a **travel graph** or
distance–time graph.

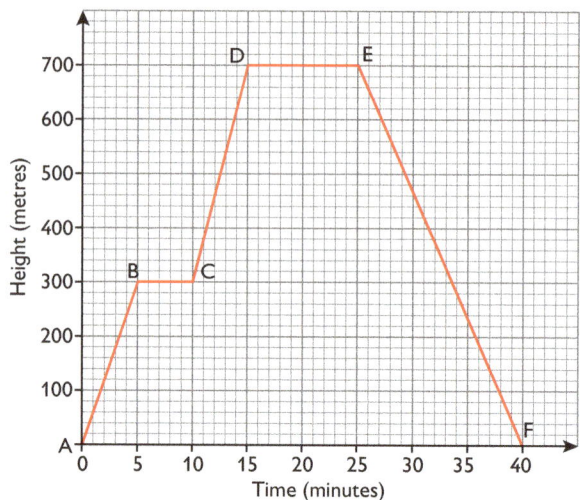

? How high does the balloon go?
What is shown by the line BC and DE?

? For how long is the balloon at a height
of 300 m?
How long does the balloon take to land?
When is the balloon at a height of 500 m?

Matt and Brad want to draw a travel graph of their journey to school.

*It takes
me 3 minutes to walk 200 m
to the corner of my road. I wait
there for 4 minutes for
Brad.*

*It then takes
us 10 minutes to walk 600 m
to the shop. We spend 5 minutes
in the shop and 2 minutes
walking 100 m to school.*

Here is Matt's travel graph.

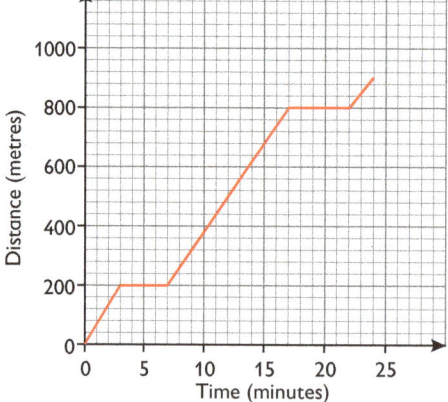

? Which part of the graph
shows Matt walking
the fastest?
How can you tell?

Task

1 Describe your journey to school.
2 Draw a travel graph of your journey.
3 Make a poster of your graph. Explain each section.
4 Make up two questions for a friend to answer about your travel graph.

Exercise

1 Mysa leaves home at 10.00 am to cycle to Avonford.
Here is a travel graph of her journey.

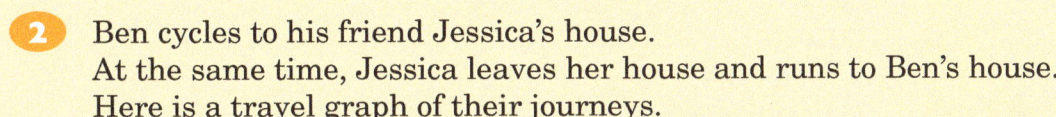

(a) How far is Avonford?

(b) How long does it take for Mysa to get to Avonford?

On the way Mysa gets a puncture.

(c) How long does it take for Mysa to mend her puncture?

Mysa spends 1 hour in Avonford.
It takes her 36 minutes to cycle home.

(d) Draw the complete travel graph for Mysa's journey.

2 Ben cycles to his friend Jessica's house.
At the same time, Jessica leaves her house and runs to Ben's house.
Here is a travel graph of their journeys.

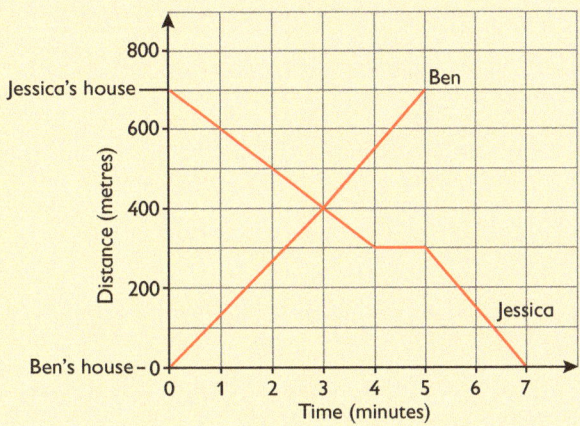

(a) After how long do the two friends pass each other?

(b) How far has Jessica travelled when she passes Ben?

(c) Who travels the faster? How can you tell?

3 The table shows Theresa's journey to work.

Time	8.00	8.10	8.20	8.25	8.30	8.40
Distance from home (kilometres)	0	6	10	10	14	20

(a) Draw a travel graph to show Theresa's journey.

(b) How far is Theresa from home at (i) 8.05 (ii) 8.35?

Finishing off

Review exercise

1 Water from a tap runs steadily into this 3 litre container.

It flows at 1 litre every 20 seconds.

Draw a graph to show the volume of water in the container against time.

2 Gary cooks some frozen peas.
He records the temperature of the water every 10 seconds after adding the peas.
Gary displays his data on a graph.

Explain the shape of Gary's graph.

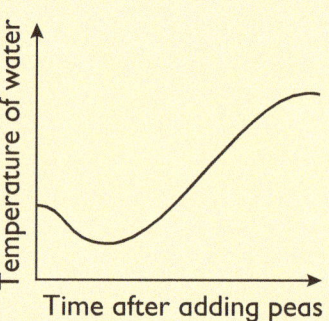

3 Yangzom hires a canoe at a boating pond.

Boat hire charges

Type of boat	Fixed charge	Rate per hour
Canoe	€4	€2
Motor boat	€2	€5
Rowing boat	€1	€3

Part hours allowed
You only pay for the time you use.
Maximum time 4 hours

(a) Draw a graph of cost against time (up to 4 hours) for a canoe.
(b) On the same graph add in lines for a motor boat and a rowing boat.
(c) Which is most expensive, a canoe, a motor boat or a rowing boat?

4 This is a conversion graph between stones and kilograms.

(a) A small child weighs two and a half stones.
What is this in kilograms?

(b) Mary weighs 55 kilograms.
What is her weight in stones?

(c) How many kilograms are there in 1 stone?

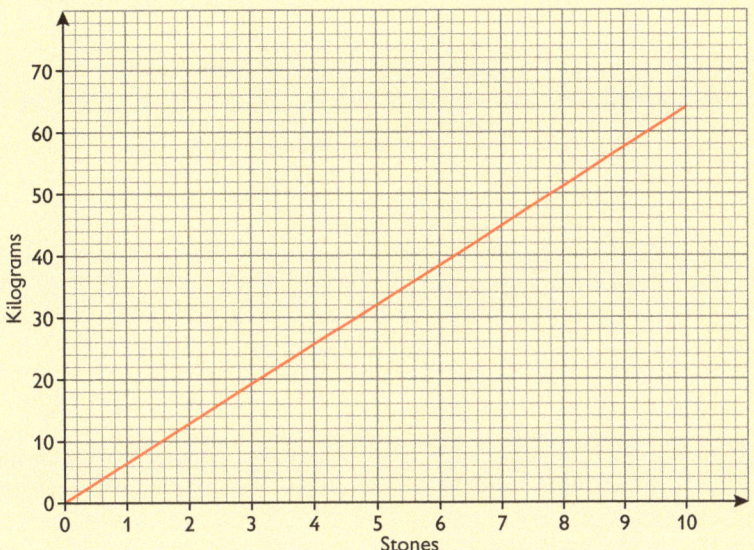

5 Paula cycles from Bridgetown to Avonford and then back again.
Here is a travel graph showing her journey.

(a) Explain the shape of
the graph.

(b) How long does Paula
spend in Avonford?

On the way Paula
gets a puncture.

(c) How far is Paula *from*
Avonford when she
has to stop?

(d) How long does Paula
spend repairing her
puncture?

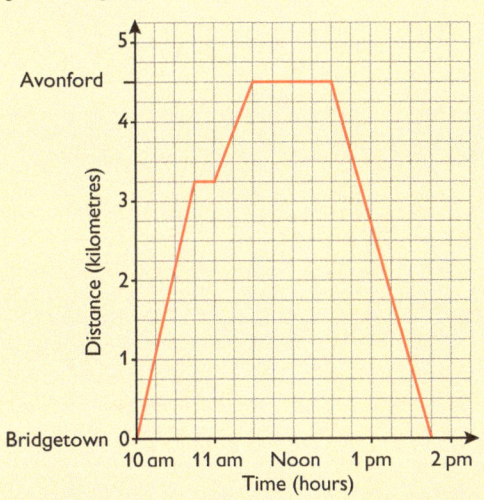

Jamie leaves Avonford at 10.30 am and walks straight to Bridgetown.
Jamie's journey takes him 45 minutes.

(e) Copy Paula's travel graph onto some graph paper.
Add a line to your graph representing Jamie's journey.

(f) At what time does Jamie pass Paula?

(g) How far is Paula *from* Bridgetown at this time?

9 Negative numbers

Directed numbers can be shown on a number line. ← *Directed numbers are positive and negative numbers.*

? Write these numbers in order, starting with the smallest.

−5, 4, −1.2, 7, 3.4, −5.1

? Write down a number that is between −1 and −1.5.
Find the number that is halfway between −2 and +4.

```
 −6   −5   −4   −3   −2   −1    0   +1   +2   +3   +4   +5   +6   +7   +8   +9
```

? Use the number line to work out the following:

(a) −2 + 3   (b) 4 + −5   (c) −1 + −5   (d) 6 − 9

Task

Class 8 are testing how well an insulating flask keeps its contents hot or cold. They measure the *outside* room temperature (A) and the temperature *inside* the flask (B) every hour.

1 Copy and complete the results table for Experiment 1.
Boiling water at 100°C is placed in the flask.

A	room temperature	18°	19°	17°	19°
B	temperature inside flask	100°	98°	97°	96.5°
A − B temperature difference					

(a) Why are all the answers for A − B negative?

(b) How much has the temperature inside the flask changed?

(c) Work out 96.5 − 100 to answer this. What does the negative answer tell you?

2 Copy and complete the results table for Experiment 2.
Ice at −10°C is placed in the flask.

A	room temperature	18°	19°	17°	19°
B	temperature inside flask	−10°	−8°	−7.5°	−6.5°
A − B temperature difference					

(a) Why are all the answers for A − B positive?

(b) Work out (−6.5) − (−10). What does the answer tell you?

? Write each of these problems as an addition and then solve it.

1 The temperature is −5°C.
What will the temperature be after a rise of 7°C?

2 A lift starts at the second floor.
It goes up 3 floors and then down 5.
Where is it now?

Exercise

1 Write the following in order of size, smallest first.

$$-2, -1.25, -1.5, -2.1, -1, -2.3$$

2 Write down a number that lies between
  (a) -1 and $+4$ **(b)** -5 and -4 **(c)** -2.1 and -2.3

3 Find the number that is exactly half way between
  (a) -3 and $+1$ **(b)** -2 and -3 **(c)** -0.5 and $+1.5$

4 Work out the following
  (a) $-7 + (-10)$ **(b)** $18 - (-35)$ **(c)** $-24 + (-19)$
  (d) $100 - (-35)$ **(e)** $0.3 + (-0.2)$ **(f)** $(-0.7) - (-0.2)$
  (g) $2.6 + (-3.5)$ **(h)** $1.75 - (-2.35)$ **(i)** $-1 + 3 + (-6) + 2$
  (j) $2 - (-7) + 4 - (-9)$ **(k)** $7 - (-5) - (+2) - (-10)$

5 For each of these problems
  (i) write down an addition
  (ii) solve the problem.
  (a) A golfer scores 2 below par (-2) on the first hole and 3 over par $(+3)$
       on the second hole.
       How many above or below par is this after the second hole?
  (b) Rob turns a wheel 30° clockwise $(+)$ and 70° anticlockwise $(-)$.
       Describe the new position of the wheel.
  (c) Mr Brown has an overdraft of €30 (i.e. $-$€30).
       How much does he have after he receives a cheque for €45?

6 For each of the following problems
  (i) write a subtraction
  (ii) solve the problem.
  (a) A diver 30 m below the surface dives another 20 m down.
       What depth is he below the surface now?
  (b) John arrived at the bus stop 20 minutes early (-20).
       The bus was 10 minutes late $(+10)$. How long did he have to wait?
  (c) Cleopatra died in 30 BC (-30).
       Rome was founded in 753 BC (-753).
       How many years before Cleopatra's death was Rome founded?

Activity

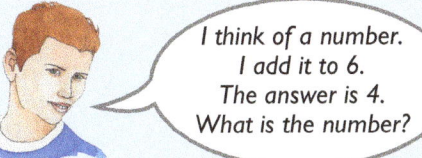

*I think of a number.
I add it to 6.
The answer is 4.
What is the number?*

*I think of a number.
I take it away from 5.
The answer is 7.
What is the number?*

1 When does a number get smaller after something has been added to it?
2 When does a number get larger after something has been subtracted from it?
3 Make up *Think of a number* questions like these to try on your friends.

Multiplying and dividing negative numbers

Task

Hassan is drawing a graph of $y = 3x$.
He has worked out the following table
of values and plotted them on a graph.

x	1	2	3
$y = 3x$	3	6	9

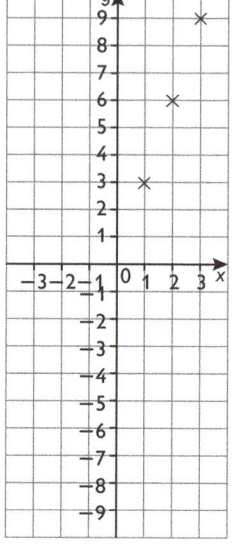

1 Draw axes showing values of x from -3 to $+3$ and y from
-10 to $+10$.

2 Plot the points on your graph and join them with a straight line.

Use your graph to copy and
complete the table.

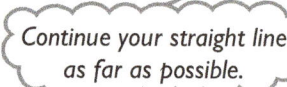

*Continue your straight line
as far as possible.*

x	-3	-2	-1	0
$y = 3x$				

For each number in the table, check that the y value is
$3 \times$ the x value.

Task

Draw the graph of $y = -2x$.
Draw axes showing values of x from
-3 to $+3$ and y from -6 to $+6$.
Plot the points in this table.

x	1	2	3
$y = -2x$	-2	-4	-6

Extend your straight line and
copy and complete this table.

x	-3	-2	-1	0
$y = -2x$				

For each of the numbers in your table check that the y number is $(-2) \times$ the value of x.

? **Work out** (a) $(-6) \div 3$ (b) $(-1.5) \div 3$ (c) $(-4.5) \div 3$.
The answers are all on your first graph. Explain.

? **Work out** (a) $4 \div (-2)$ (b) $(-6) \div (-2)$ (c) $5 \div (-2)$ (d) $(-3) \div (-2)$.
Use your second graph to check your answers.

? **Fill in these tables.**
They help you work out when an answer is negative
or positive when you multiply or divide directed
numbers.

$\times$	$+$	$-$
$+$		
$-$		

$\div$	$+$	$-$
$+$		
$-$		

Exercise

1 Evaluate

(a) $4 \times (-5)$ **(b)** $(-10) \times 30$ **(c)** $(-14) \times (-20)$

(d) $(-5)^2$ **(e)** $200 \times (-40)$ **(f)** $(-500) \times (-300)$

(g) $(-0.4) \times 0.7$ **(h)** $(-0.3)^2$ **(i)** $(-1.2)^2$

2 Work out

(a) $12 \div (-6)$ **(b)** $(-87) \div 3$ **(c)** $(35) \div (-14)$

(d) $-16 \div (-32)$ **(e)** $400 \div (-20)$ **(f)** $(-120) \div (-30)$

(g) $0.6 \div (-0.02)$ **(h)** $(-18) \div 0.36$ **(i)** $(-5.2) \div (-1.3)$

3 Copy and complete the following multiplication table.

×	6				12
5					
4		−16			
−2					
−3			9		
−10				70	

4 Complete the following table of values for the graph of $y = -4x$. Draw the graph.

x	−3	−1	1	3
$y = -4x$				

5 George multiplied two whole numbers together and got the answer -18.
Write down three pairs of numbers that he could have used.

6 The result of multiplying two numbers is 24.
Write down all the pairs of whole numbers that give this result.

7 **(a)** Work out 3×1.5, 3×0.5, $3 \times (-0.5)$, $3 \times (-1.5)$, $3 \times (-2.5)$

     (b) Use your first graph from the Task to check your answers.

     (c) Work out $(-2) \times 0.5$, $(-2) \times (-0.5)$, $(-2) \times (-2.5)$

     (d) Use your second graph from the Task to check your answers.

Investigation

$(-2) \times (-3) \times (-4) = (+6) \times (-4) = -24$ OR $(-2) \times (-3) \times (-4) = (-2) \times (+12) = -24$
Multiplying three negative numbers together has given a negative answer.

? **Is this always true? Try some more examples of your own.**

1 Work out   **(a)** $(+2) \times (-3) \times (-4)$ **(b)** $(-2) \times (+3) \times (-4)$ **(c)** $(-2) \times (-3) \times (+4)$

2 Work out   **(a)** $(+2) \times (+3) \times (-4)$ **(b)** $(+2) \times (-3) \times (+4)$ **(c)** $(-2) \times (+3) \times (+4)$

Explain how to decide whether the answer will be positive or negative when multiplying 3 or more directed numbers.

Finishing off

Review exercise

1 Write down three numbers that lie between -7 and 2.

2 Find the number that is exactly half way between the following pairs of numbers.

(a) -1 and -2 **(b)** -4 and -10 **(c)** -7 and 3

(d) -12 and 5   **(e)** -3.5 and -4.5 **(f)** -6.3 and -6.4

(g) -2.5 and 1.5   **(h)** -1.7 and 2.9

3 You can use the formula $\dfrac{A+B}{2}$. It gives the number that is exactly half way between the numbers A and B. Copy and complete the table. Compare your answers with Question 2.

A	B	A + B	$\dfrac{A+B}{2}$
−1.0	−2.0		
−4.0	−10.0		
−7.0	3.0		
−12.0	5.0		
−3.5	−4.5		
−6.3	−6.4		
−2.5	1.5		
−1.7	2.9		

4 Work out the following

(a) $7 + (-3)$ **(b)** $(-10) + (-24)$

(c) $(-1.3) + 2.9$ **(d)** $(-16) + (+12) + (-8) + (+14)$

(e) $(-6) + (-2) - (-3)$ **(f)** $(-40) - (+20) + (-30)$

(g) $4.7 - (-1.4)$ **(h)** $8.2 + (-4.9) - (-6.4)$

(i) $(-2) \times (-3) \times (-4)$ **(j)** $(-5) \times (+6) \times (-2)$

(k) $(-1) \times (-1) \times (-1) \times (-1)$ **(l)** $(-1)^5$ **(m)** $18 \div (-3)$

(n) $(-27) \div (-9)$ **(o)** $4.2 \div (-1.4)$ **(p)** $(-1.44) \div (+1.2)$

5 Each question in an examination has four possible answers.
Only one answer is correct.
A candidate receives 5 marks for a correct answer and (-2) marks for a wrong answer.

M is the total mark.

(a) Explain the formula $M = 5C - 2W$.

W is the number of wrong answers.

C is the number of correct answers.

(b) Copy and complete the table to show the marks obtained by the following candidates.

	C	W	$5C$	$-2W$	$5C - 2W$
Jane	10	10			
Edward	12	8			
Jessica	9	11			
Davinda	5	15			

(c) George knows he chose the correct answer for 8 of the questions and one of his answers is wrong.
He guessed the other 11 answers.

(i) What is the highest mark that he can get?

(ii) What is the lowest mark?

6 The formula $v = u + at$ is often used in physics.
Use the formula to find v when

(a) $u = 2$, $a = 5$ and $t = 6$

(b) $u = 4$, $a = -10$ and $t = 3$

(c) $u = -30$, $a = 0.5$ and $t = 6$

(d) $u = 3.5$, $a = 1.5$ and $t = -4$

Activity

1 A safe can be opened by successive turns of a number dial.
The number 0 is at the top.
The sequence of turns to open the safe is described by the following calculation.

$(+2) + (-4) + (+6) + (-2) + (-3)$

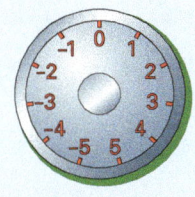

Turn 2 to the left then 4 to the right …

(a) Which number is at the top after each turn of the dial?

(b) Which number is at the top when the safe is opened?

2 Jack the burglar is given the code for a different safe.

$2, -1, 5, 1, -3, -5$

These numbers must appear at the top after each turn.
Jack works in the dark and cannot read the numbers.
Help him decide the turns he needs to open the safe.

3 Write down and describe your own safe code.
Ask a friend to find the turns needed to crack your code.

10 Fractions

Equivalent fractions

Here are Jane's last three maths marks.

Jane's mother wants to know if she is improving.

Her mark for Test 1 is written as a fraction $\frac{15}{25}$.

Equivalent fractions can be found by
multiplying or dividing both top and bottom lines.

Test 1 Jane Smith
$\frac{15}{25}$

Test 2 Jane Smith
$\frac{12}{20}$

Test 3 Jane Smith
$\frac{45}{60}$

Is your Maths getting any better Jane?

$$\frac{15}{25} \underset{\substack{\text{divide} \\ \text{by 5}}}{=} \frac{3}{5} \underset{\substack{\text{multiply} \\ \text{by 20}}}{=} \frac{60}{100} \underset{\substack{\text{divide} \\ \text{by 2}}}{=} \frac{30}{50}$$

 Write down two more fractions that are equivalent to $\frac{15}{25}$.

When the bottom line is 100 the fraction is a percentage.

This is called the denominator.

$\frac{60}{100}$ is '60 out of 100'. This is 60%.

Percentage means 'out of 100'.

Task

1 Jane's mark for Test 2 is written as a fraction.
   Show that it is equivalent to her mark for Test 1.

   This gives a percentage.

2 Write her mark for Test 3 as an equivalent fraction
   (a) with a denominator of 20   **(b)** with a denominator of 100.

3 What percentage did Jane obtain for Test 3? Has she improved?

 How do the equivalent fractions in Question 2 help you answer this?

Common denominators

To compare fractions they need the same bottom line.
This is called the *common denominator*.

 Write each of the fractions $\frac{3}{4}$, $\frac{2}{5}$ and $\frac{5}{8}$ with a bottom line of 40.
Use this to write the fractions in order, smallest first.

 Find a suitable common denominator to compare the fractions $\frac{1}{2}$, $\frac{3}{5}$ and $\frac{7}{10}$.
Write each of these fractions as a percentage.
Which is easier: comparing fractions with the same bottom line or comparing percentages?

Exercise

1 Which of these fractions are equivalent to $\frac{4}{5}$?

(a) $\frac{8}{10}$ (b) $\frac{5}{6}$ (c) $\frac{12}{15}$ (d) $\frac{8}{9}$ (e) $\frac{40}{50}$ (f) $\frac{80}{100}$

(g) $\frac{16}{25}$ (h) $\frac{44}{55}$ (i) $\frac{14}{15}$ (j) $\frac{34}{35}$ (k) $\frac{404}{505}$ (l) $\frac{5}{4}$

2 Which of these fractions are equivalent to $\frac{9}{12}$?

(a) $\frac{18}{24}$ (b) $\frac{6}{9}$ (c) $\frac{12}{15}$ (d) $\frac{12}{16}$ (e) $\frac{15}{20}$ (f) $\frac{30}{40}$

3 Find five pairs of equivalent fractions from the list below.

(a) $\frac{3}{4}$ (b) $\frac{4}{7}$ (c) $\frac{3}{8}$ (d) $\frac{6}{10}$ (e) $\frac{3}{9}$ (f) $\frac{2}{8}$

(g) $\frac{3}{5}$ (h) $\frac{6}{16}$ (i) $\frac{3}{12}$ (j) $\frac{9}{12}$ (k) $\frac{8}{14}$ (l) $\frac{4}{12}$

4 **(i)** Write the following fractions with a common denominator.
(ii) Use this to write the fractions in order starting with the smallest.

(a) $\frac{1}{2}$ (b) $\frac{2}{3}$ (c) $\frac{7}{12}$ (d) $\frac{3}{4}$ (e) $\frac{11}{24}$

5 **(i)** Write the following fractions as percentages.
(ii) Place them in order starting with the largest.

(a) $\frac{1}{2}$ (b) $\frac{2}{5}$ (c) $\frac{3}{10}$ (d) $\frac{7}{20}$ (e) $\frac{8}{25}$

6 What is Jane's best test mark from the previous page?
Her next test is to be marked out of 50. She wants to get a higher score.
How many marks must she get?

7

Write 8 out of 10 as **(i)** a fraction
(ii) a percentage.

In a group of 30 cats how many would you expect to prefer Kati-kins?

Activity Make a collection of newspaper headlines and advertisements that make claims like the one in Question 7.
Display these on a poster.
Use equivalent fractions and percentages to explain each statement.

Adding fractions

Daniel's mother has two partly used jars of coffee.

She puts all of the coffee into one jar.

Daniel works this out for his mother.
He changes both fractions into $\frac{1}{12}$ ths.

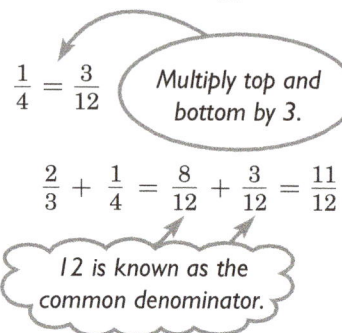

$\frac{1}{4} = \frac{3}{12}$ *Multiply top and bottom by 3.*

Multiply top and bottom by 4. $\frac{2}{3} = \frac{8}{12}$

$\frac{2}{3} + \frac{1}{4} = \frac{8}{12} + \frac{3}{12} = \frac{11}{12}$

The answer can be found by adding the top numbers (numerators).

12 is known as the common denominator.

The jar is $\frac{11}{12}$ ths full.

? 8, 16 and 32 could all be chosen as a common denominator for $\frac{1}{4}$ and $\frac{3}{8}$.

8 is called the **Lowest Common Denominator**.

Why is it easier to work with the Lowest Common Denominator?
Change $\frac{1}{4}$ into $\frac{1}{8}$ ths. Work out $\frac{1}{4} + \frac{3}{8}$.

Task

1 For each of the following pairs of fractions:

 (a) find the lowest common denominator   **(b)** add the fractions

  (i) $\frac{2}{5}$ and $\frac{1}{5}$ **(ii)** $\frac{3}{4}$ and $\frac{1}{8}$ **(iii)** $\frac{1}{3}$ and $\frac{1}{5}$

  (iv) $\frac{1}{2}$ and $\frac{2}{7}$ **(v)** $\frac{5}{16}$ and $\frac{1}{4}$ **(vi)** $\frac{1}{6}$ and $\frac{3}{8}$

2 Fill in the missing numbers in each of the following:

  (i) $\frac{1}{2} + \frac{1}{4} + \frac{1}{8} = \frac{?}{8}$ **(ii)** $\frac{1}{2} + ? = \frac{3}{4}$ **(iii)** $? + \frac{5}{7} = \frac{13}{14}$ **(iv)** $\frac{?}{6} + \frac{?}{8} = \frac{13}{?}$

Fractions are subtracted in the same way.

Example

$$\frac{2}{5} - \frac{1}{4} = \frac{8}{20} - \frac{5}{20} = \frac{3}{20}$$

Subtract the top numbers.

20 is the common denominator.

? What happens when you add $\frac{3}{20}$ and $\frac{1}{4}$?
How does this check your answer?

 **?** Work out **(a)** $\frac{3}{5} - \frac{3}{10}$ **(b)** $\frac{2}{3} - \frac{1}{7}$.
How can you check your answers?

Exercise

1 Work out the following:

(a) $\frac{1}{7} + \frac{2}{7}$ (b) $\frac{1}{4} + \frac{3}{4}$ (c) $\frac{4}{5} - \frac{3}{5}$ (d) $\frac{5}{8} - \frac{3}{8}$

2 (a) Find the smallest number that is a multiple of 2 and 3.
This is known as the Lowest Common Multiple (LCM) of 2 and 3.

(b) Find the LCM of the following pairs of numbers:
(i) 4 and 5   (ii) 3 and 6   (iii) 4 and 6   (iv) 5 and 7   (v) 9 and 6.

3 Use the answers to question 2 to help you to work out the following:

(a) $\frac{3}{4} + \frac{1}{5}$ (b) $\frac{5}{6} - \frac{1}{3}$ (c) $\frac{3}{4} + \frac{1}{6}$ (d) $\frac{2}{5} + \frac{3}{7}$ (e) $\frac{7}{9} - \frac{1}{6}$

4 It takes $\frac{1}{3}$ of a tin to paint the hall and $\frac{2}{5}$ of a tin for the bathroom.
How much paint will be left out of a full tin after painting both?

5 Can all the shampoo be put
into one bottle?

6 Calculate the following:

(a) $\frac{5}{12} + \frac{7}{12}$ (b) $\frac{10}{19} - \frac{3}{19}$ (c) $\frac{3}{4} - \frac{5}{12}$ (d) $\frac{2}{5} + \frac{3}{10}$

(e) $1 - \frac{2}{5}$ (f) $\left(\frac{3}{7} + \frac{1}{14}\right) - \frac{2}{7}$ (g) $\frac{2}{5} - \frac{1}{10} + \frac{3}{5}$ (h) $\frac{1}{3} + \frac{1}{4} + \frac{1}{5}$

Investigation   A reservoir is $\frac{3}{4}$ full of water.
Each day $\frac{1}{20}$ th of its full capacity is used to supply the local village.
On a hot day $\frac{1}{100}$ th is lost through evaporation.
When it rains heavily the reservoir receives $\frac{1}{5}$ th of its capacity in
a day.

Here is the weather report for a week in August.

Monday	Hot and sunny
Tuesday	Heavy rain
Wednesday	Heavy rain
Thursday	Cloudy and cool
Friday–Sunday	Hot and sunny

Explain why the calculation $\frac{3}{4} - \frac{1}{20} - \frac{1}{100}$ will give the fraction of
water remaining in the reservoir after Monday.

How much will remain (i)   after Tuesday
(ii)   at the end of the week?

Mixed numbers

? **Can all the sand be placed in one bucket?**

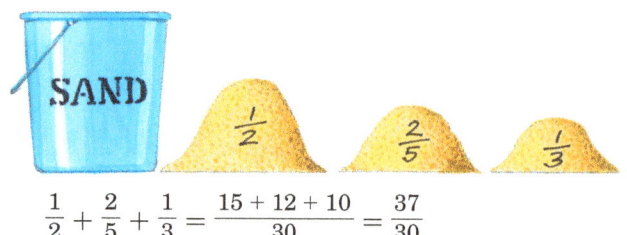

$$\frac{1}{2} + \frac{2}{5} + \frac{1}{3} = \frac{15 + 12 + 10}{30} = \frac{37}{30}$$

$\frac{30}{30}$ is one whole 1

$\frac{37}{30}$ gives one whole 1 and $\frac{7}{30}$.

That is $1\frac{7}{30}$.

$\frac{37}{30}$ is called a *top-heavy fraction* $1\frac{7}{30}$ is a *mixed number*.

Task

1 Copy this into your books and fill in the blank spaces.

$$1 = \frac{}{3} = \frac{5}{} = \frac{}{17} = \frac{21}{}$$

2 Change these top-heavy fractions into mixed numbers.

$$\frac{3}{2}, \quad \frac{6}{5}, \quad \frac{13}{10}, \quad \frac{25}{21}, \quad \frac{42}{37}$$

3 2 can be written as $\frac{2}{1}$, as $\frac{6}{3}$ or as $\frac{10}{5}$. Write down three more fractions that are equivalent to 2.

4 These top-heavy fractions are all larger than 2. Write them as mixed numbers.

$$\frac{7}{3}, \quad \frac{15}{6}, \quad \frac{21}{10}, \quad \frac{30}{12}$$

5 Explain why the top-heavy fraction $\frac{38}{9}$ can be written as the mixed number $4\frac{2}{9}$.

6 Write these top-heavy fractions as mixed numbers.

$$\frac{9}{4}, \quad \frac{12}{5}, \quad \frac{19}{11}, \quad \frac{65}{31}, \quad \frac{70}{13}$$

? **George needs $\frac{3}{4}$ of a tin of paint to paint his bedroom.**
What fraction will he have left?

He works out $\quad 1\frac{1}{3} - \frac{3}{4}$

$$\frac{4}{3} - \frac{3}{4} = \frac{16 - 9}{12} = \frac{7}{12}$$

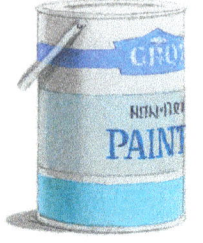

Full size

$\frac{1}{3}$ *size*

? **Write $1\frac{2}{5}$ as a top-heavy fraction.**
Use this to work out $1\frac{2}{5} - \frac{2}{3}$.

Exercise

1 State **(i)** the mixed number and **(ii)** the top-heavy fraction that is shaded in each of the diagrams below.

(a)

(b)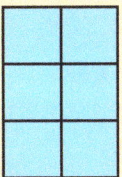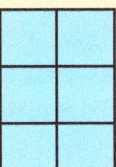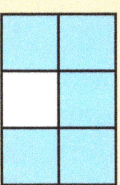

2 Copy the following diagrams and shade in the given fraction.

(a)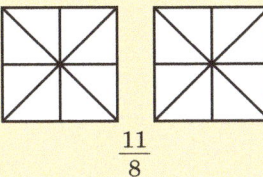

$\frac{11}{8}$

(b)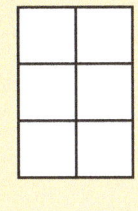

$\frac{15}{6}$

3 Write the number 3 as a fraction in five different ways.

4 Change these top-heavy fractions to mixed numbers

(a) $\frac{3}{2}$ **(b)** $\frac{5}{3}$ **(c)** $\frac{5}{2}$ **(d)** $\frac{7}{3}$ **(e)** $\frac{7}{4}$ **(f)** $\frac{9}{5}$ **(g)** $\frac{11}{5}$

5 Change these mixed numbers to top-heavy fractions

(a) $1\frac{1}{2}$ **(b)** $1\frac{1}{4}$ **(c)** $1\frac{2}{3}$ **(d)** $1\frac{4}{5}$

(e) $1\frac{2}{7}$ **(f)** $2\frac{1}{2}$ **(g)** $2\frac{1}{3}$ **(h)** $3\frac{1}{2}$

(i) $4\frac{1}{2}$ **(j)** $5\frac{2}{3}$

6 Work out the following, giving your answers as mixed numbers

(a) $\frac{2}{5}+\frac{4}{5}$ **(b)** $\frac{1}{6}+\frac{5}{6}+\frac{5}{6}$ **(c)** $\frac{3}{11}+\frac{5}{11}+\frac{6}{11}$

(d) $\frac{1}{7}+\frac{2}{7}+\frac{4}{7}$ **(e)** $\frac{4}{5}+\frac{7}{10}$ **(f)** $\frac{3}{4}+\frac{7}{12}$

(g) $\frac{7}{9}+\frac{2}{3}$ **(h)** $\frac{5}{6}+\frac{5}{8}$ **(i)** $\frac{1}{2}+\frac{3}{4}+\frac{1}{8}$

(j) $\frac{2}{3}+\frac{5}{8}+\frac{11}{12}$ **(k)** $\frac{1}{2}+\frac{1}{3}+\frac{1}{4}+\frac{1}{5}$

7 Work out the following

(a) $1-\frac{7}{9}$ **(b)** $1\frac{1}{2}-\frac{3}{4}$ **(c)** $1\frac{2}{5}-\frac{4}{5}$ **(d)** $1\frac{5}{7}-\frac{6}{7}$

(e) $1\frac{3}{4}-\frac{7}{8}$ **(f)** $1\frac{2}{3}-\frac{5}{6}$ **(g)** $1\frac{5}{8}-\frac{9}{16}$ **(h)** $1\frac{3}{10}-\frac{4}{5}$

Finishing off

Now that you have finished you should be able to:

- find equivalent fractions
- write fractions as percentages
- use equivalent fractions or percentages to compare the size of fractions
- add and subtract fractions
- change top-heavy fractions into mixed numbers.

Review exercise

1 Write down sets of equivalent fractions from the list below.

$$\frac{3}{4} \qquad \frac{5}{6} \qquad \frac{3}{7} \qquad \frac{6}{8} \qquad \frac{4}{5} \qquad \frac{10}{12}$$

$$\frac{9}{21} \qquad \frac{40}{50} \qquad \frac{1}{3} \qquad \frac{2}{4} \qquad \frac{15}{18}$$

$$\frac{4}{12} \qquad \frac{5}{10} \qquad \frac{50}{60} \qquad \frac{25}{50} \qquad \frac{13}{39} \qquad \frac{16}{20}$$

2 Write the fractions in each pair with a common denominator.
Use this to help you to choose the larger fraction from each pair.

(a) $\frac{1}{4}$ and $\frac{3}{8}$ (b) $\frac{3}{5}$ and $\frac{4}{7}$ (c) $\frac{3}{4}$ and $\frac{5}{6}$ (d) $\frac{5}{9}$ and $\frac{2}{3}$

(e) $\frac{2}{5}$ and $\frac{4}{15}$ (f) $\frac{2}{3}$ and $\frac{4}{5}$ (g) $\frac{5}{12}$ and $\frac{7}{18}$ (h) $\frac{1}{6}$ and $\frac{2}{15}$

3 Work out the following:

(a) $\frac{3}{7} + \frac{4}{9}$ (b) $\frac{5}{6} - \frac{2}{3}$ (c) $\frac{1}{5} + \frac{1}{10}$ (d) $\frac{7}{9} - \frac{5}{18}$

(e) $\frac{1}{6} + \frac{1}{4}$ (f) $\frac{3}{8} + \frac{5}{12}$ (g) $\frac{2}{3} - \frac{2}{7}$ (h) $\frac{7}{12} - \frac{3}{8}$

(i) $\frac{1}{2} + \frac{1}{4} - \frac{5}{8}$ (j) $\frac{1}{4} - \frac{9}{16} + \frac{3}{8}$ (k) $\frac{1}{6} - \frac{1}{3} + \frac{1}{2}$ (l) $\frac{3}{10} - \frac{1}{2} + \frac{2}{5}$

4 Work out the following.
Give your answers as mixed numbers.

(a) $\frac{4}{5} + \frac{3}{5}$ (b) $\frac{5}{6} + \frac{3}{8}$ (c) $\frac{2}{5} + \frac{2}{3}$ (d) $\frac{3}{4} + \frac{7}{12}$

(e) $\frac{5}{6} + \frac{5}{8}$ (f) $\frac{5}{6} + \frac{7}{8}$ (g) $\frac{8}{9} + \frac{2}{3}$ (h) $\frac{1}{2} + \frac{1}{3} + \frac{1}{4}$

5 Write the following fractions as percentages:

(a) $\frac{1}{2}$ (b) $\frac{3}{4}$ (c) $\frac{1}{10}$ (d) $\frac{3}{10}$ (e) $\frac{1}{20}$ (f) $\frac{7}{20}$

(g) $\frac{1}{5}$ (h) $\frac{2}{5}$ (i) $\frac{1}{25}$ (j) $\frac{9}{25}$ (k) $\frac{1}{50}$ (l) $\frac{3}{50}$

6 Rachel scores 3 out of 7. Charles scores 4 out of 9.
Whose score is better? Explain your answer.

7 The table below shows the activities chosen by 75 children in year 8.
Copy and complete the table.

Activity	Number of children	As a fraction	As a percentage
Squash	9	$\frac{9}{75} = \frac{3}{25}$	12%
Computer games	18		
Tennis	15		
Orchestra	27		
Chess	6		
	Total 75	Total 1	Total 100%

8 The pie chart shows the colours in a box of sweets.
(a) An angle of 30° is used for the blue sweets.
This shows that the fraction of blue sweets
is $\frac{30}{360} = \frac{1}{12}$.

What fraction of the sweets are **(i)** orange
(ii) green
(iii) pink?

(b) There are 120 sweets in the box.

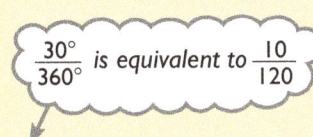

$\frac{30°}{360°}$ is equivalent to $\frac{10}{120}$

There are 10 blue sweets in the box.

Work out the number of each colour of sweets in the box, showing your
working in the same way as given above..

9 Work out the following.
The answers are all mixed numbers.

(a) $\frac{3}{4} + \frac{1}{2}$ **(b)** $\frac{3}{4} + \frac{1}{3} + \frac{5}{6}$ **(c)** $\frac{8}{9} + \frac{2}{3} + \frac{4}{5}$ **(d)** $\frac{9}{10} + \frac{7}{15} + \frac{11}{20}$

(e) $\frac{1}{2} + \frac{2}{3} + \frac{3}{4} + \frac{4}{5}$ **(f)** $\frac{5}{8} + \frac{9}{16} - \frac{1}{32}$ **(g)** $\frac{3}{4} + \frac{1}{2} - \frac{1}{5}$ **(h)** $\frac{7}{8} + \frac{5}{16} + \frac{7}{16} - \frac{1}{4}$

10 Find a fraction that lies between each of the following pairs of fractions.

Example $\frac{1}{2}$ and $\frac{3}{4}$.

$\frac{1}{2} = \frac{4}{8}$, $\frac{3}{4} = \frac{6}{8}$ so $\frac{5}{8}$ lies between $\frac{1}{2}$ and $\frac{3}{4}$.

(a) $\frac{1}{2}$ and $\frac{1}{4}$ **(b)** $\frac{1}{2}$ and $\frac{1}{3}$ **(c)** $\frac{1}{2}$ and $\frac{2}{3}$

(d) $\frac{1}{4}$ and $\frac{2}{5}$ **(e)** $\frac{3}{8}$ and $\frac{3}{4}$

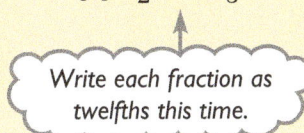

Write each fraction as
twelfths this time.

11 Converting units

Measures

The original units of length came from different parts of the human body.

A foot was the length of a fully grown man's foot.

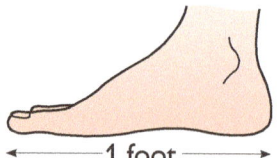

1 foot

Task

1 Measure the distance across your classroom putting your feet like this and counting.

How many feet across is the room?

2 Now measure it in paces. 1 pace is about 1 yard or 3 feet.

? **Do your measurements agree?**

The Romans decided that an inch was one-twelfth of a foot.

*The word **inch** comes from the Latin **unciae**, meaning twelfths.*

? **Halley is 4 feet and 11 inches tall.
How many inches is this?**

? **Lois is 66 inches tall.
What is this in feet and inches?**

? **The distance from the end of a person's thumb to the knuckle is about one inch.
What is this distance on your thumb?
What is the average for your class?**

? **Here are various imperial units conversions, but some of the numbers and words are missing.
What should they be?**

Length	Mass (weight)	Capacity
12 inches = 1 foot	16 ounces (oz) = 1 pound (lb)	8 pints = 1 ......
3 feet = 1 yard	14 pounds = 1	
1760 yards = 1	2240 pounds = 1	

Exercise

1 The heights of a family are given in feet and inches. Write them in inches.

(a) Dad   6 ft 3 ins   **(b)** Mum   5 ft 5 ins

(c) Tom   4 ft 7 ins   **(d)** Emily   3 ft 8 ins

(e) James   1 ft 10 ins   **(f)** Lucy   2 ft 9 ins

2 How many feet are there in these measurements?

(a) A room 120 ins high.

(b) A cliff 300 ins high.

(c) A pterodactyl with a wing span of 276 ins.

3 A **yard** comes from the Saxon word **gyrd**, meaning girth, or the distance around someone's waist.
1 yard is 3 feet.

(a) Change these measurements to yards.
  (i) a cricket pitch of 66 ft long
  (ii) a garden 108 ft long
  (iii) the 15 ft of material needed for a dress

(b) Change these lengths to feet.
  (i) a Cetisaurus dinosaur 20 yards long
  (ii) an Ichthyosaur (marine reptile)
        3 yds long
  (iii) a Diplodocus 29 yds long

4 1 metre is about 39 inches.

(a) Is a metre longer than a yard? Explain your answer.

A mile is 1760 yards.

(b) (i) How many inches are there in a mile?
  (ii) How many inches are there in 1500 metres?
  (iii) Which race is longer, the 1500 metres or the mile?

5 Copy and complete the cross-number below.

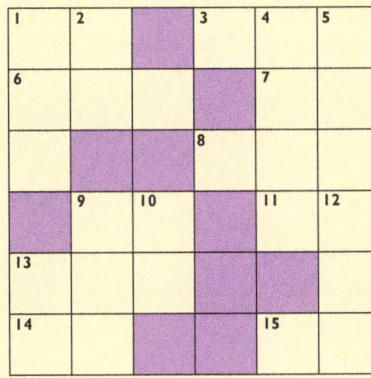

Across
1 Pints in 9 gallons
3 Inches in 71 ft
6 Inches in 15 yds
7 Gallons in 480 pints
8 Inches in 11 yds 1 ft 9 ins
9 5ft 9ins in inches
11 2 pounds less than
     5 stones
13 Pounds in 24 stones
14 Ounces in $2\frac{1}{2}$ pounds
15 3 pints over 10 gallons

Down
1 Feet in 250 yds
2 Pints in 3 gallons
4 156 yds in inches
5 Feet in 69 271 yds
9 Inches in $52\frac{1}{2}$ ft
10 Ounces in 6 pounds
13 Pounds in 2 stone
     6 pounds

Conversion graphs

Mike and Karen are driving to the South of France for a holiday.
They see this sign on the motorway.

| Lyons | 60 |
| Valence | 155 |

In France and the rest of Europe, distances are measured in kilometres.
In Britain they use miles. Kilometres are a metric unit, miles are an imperial unit.

 Think of some other metric and imperial measurements.
Which imperial units are still commonly used in Britain?

Mike and Karen want to know these distances in miles.
They draw this conversion graph to change between miles and kilometres.

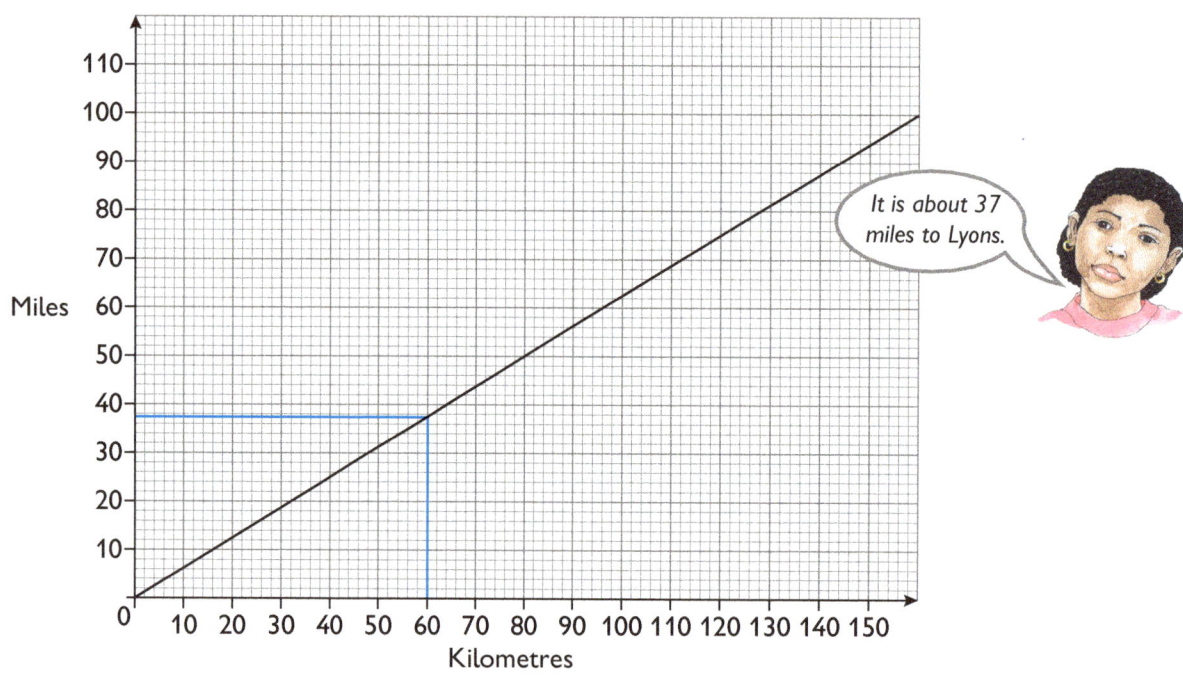

Mike and Karen have used the approximate conversion 8 km for 5 miles.

 How can you tell?
Is their graph absolutely accurate?

 Use the conversion graph to find how far it is to Valence in miles.
How would you use the graph to change miles to kilometres?

Task

During their holiday, Mike and Karen
keep a record of how many miles they
drive each day.

Monday	28	Thursday	51
Tuesday	75	Friday	97
Wednesday	34	Saturday	82

1 Use the graph to work out how far
   they drive each day in kilometres.

2 They spend a total of €420 on petrol.
   Petrol costs €0.62 per litre. Work out how many kilometres their car does to the litre.

Exercise

1 Use the conversion graph opposite to change each of these distances to miles.
(a) 150 km   **(b)** 80 km   **(c)** 95 km   **(d)** 125 km   **(e)** 44 km

2 Use the conversion graph opposite to change each of these distances to kilometres.
(a) 70 miles   **(b)** 50 miles   **(c)** 95 miles   **(d)** 98 miles   **(e)** 63 miles

3 This conversion graph is for changing between pounds and kilograms.

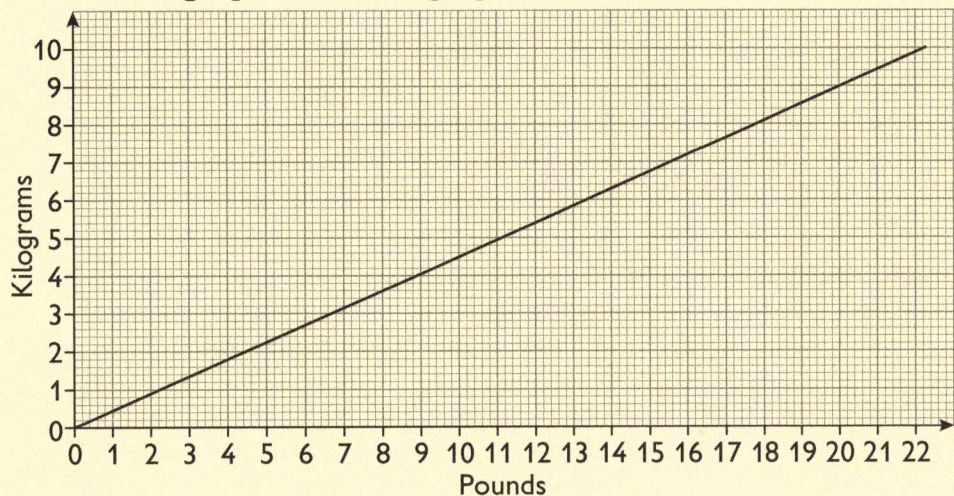

Use the graph to answer the following questions.
(a) A new-born baby weighs 3.2 kg. What is this in pounds?
(b) A turkey is labelled 5 kg. What is this in pounds?
(c) Jane wants to buy 8 pounds of potatoes. What is this in kg?
(d) A baby's car seat is suitable for weights of up to 20 pounds. What is this in kg?

4 Use the conversion graph opposite to answer the following questions.
(a) Convert to kilometres
    (i) Barnstaple to Bristol, 100 miles **(ii)** Totnes to Tavistock, 28 miles
(b) Convert to miles
    (i) Ledbury to Malvern, 16 km **(ii)** York to Kendal, 144 km

5 This graph converts between pints and litres.
Gallons are also marked on it.
(a) Convert
    (i) 36 pints into litres
    (ii) 15 litres into pints
    (iii) 18 pints into litres
    (iv) 5 litres into pints
(b) (i) How many gallons are the same as 40 pints?
    (ii) How many pints are 1 gallon?

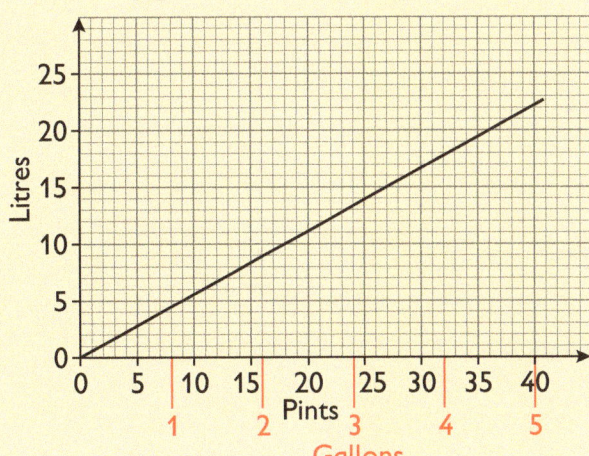

Drawing a conversion graph

Mike and Karen go shopping during their holiday in Hong Kong.

They want to know the cost of each item in euro and cent.

The exchange rate is €1 = $10.6

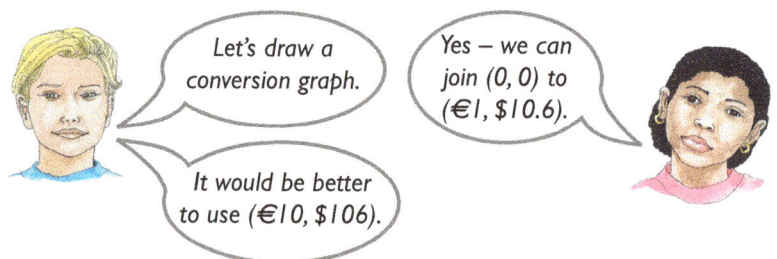

? **Why is Mike's suggestion better than Karen's?**

Task

Draw an accurate conversion graph for Hong Kong dollars against euro.

Use the graph to find the cost of the wine, chocolates and perfume in the picture in euro and cent.

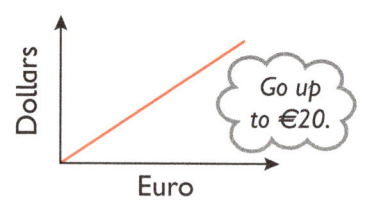

? **Work out how many Hong Kong dollars are equivalent to €20.**
Check this point is on the line.

? **Mike sees a camera he would like to buy. It costs $800.**
How can they use a conversion graph to work out how much this is in euro?
How accurate do you think this is?

? **How can you use a calculator to convert between euro and Hong Kong dollars?**

? **Would this conversion graph be useful for *any* trip to Hong Kong?**

Exercise

1 Use your conversion graph from the Task to change
   (a) each of these amounts to Hong Kong dollars.
      (i) €6   **(ii)** €18   **(iii)** €12.50   **(iv)** €3.50
   (b) each of these amounts to euro.
      (i) $150   **(ii)** $80   **(iii)** $25   **(iv)** $115

2 Simon is going on a business trip in London.
   He is using pounds sterling to pay for all his expenses.
   The exchange rate is £1 = €1.6.
   (a) Draw a conversion table for pounds and euro.
      Your graph should go up to €500.
   (b) Simon's hotel bill is £340.
      Use your graph to find out what this is in euro.
   (c) The cost of Simon's car hire is 200 pounds.
      Use your graph to find this in euro.
   (d) Simon's company has allowed €240 for buying petrol.
      Use your graph to find this in pounds.

3 Petrol is sold in litres. *A metric measure*
   It used to be sold in gallons. *An imperial measure*
   1 gallon ≈ 4.5 litres.

   The sign ≈ means 'approximately equals.

   (a) Draw a conversion graph for gallons and litres.
      Your graph should go up to 10 gallons.
   (b) Becky buys 38 litres of petrol.
      Use your graph to find out how many gallons this is.
   (c) Chris has an old car. Its manual says that the petrol tank holds
      7.5 gallons of petrol.
      Use your graph to convert this to litres.
   (d) 1 gallon is equal to 8 pints.
      Use this information, together with your graph, to find out how many
      litres there are in a 6 pint carton of milk.

4 One day the exchange rates are   €1 = 1.6 Canadian dollars
                                        €1 = 10.6 South African rand
   (a) Draw a conversion graph between Canadian dollars and
      South African rand.
   (b) Use your graph to change
      (i) 150 dollars to rand.   **(ii)** 270 dollars to rand.
      (iii) 800 rand to dollars.   **(iv)** 560 rand to dollars.

5 You can answer Question 4 without drawing a conversion graph. How?
   Use your method to work out the answers to Question 4 again.
   Which method is **(a)** more accurate?   **(b)** easier to use?

Finishing off

Now that you have finished this chapter you should be able to:

- understand imperial measures
- convert from imperial units to metric units and vice versa
- use a conversion graph
- draw a conversion graph
- make estimates.

Review exercise

Measures	Imperial	Metric	Approximate conversions
Lengths	12 inches = 1 foot 3 feet = 1 yard 1760 yards = 1 mile	10 mm = 1 centimetre 100 centimetres = 1 metre 1000 m = 1 km	$2\frac{1}{2}$ cm ≈ 1 inch 1 metre ≈ 39 inches 8 kilometres ≈ 5 miles
Mass	16 ounces = 1 pound 14 pounds = 1 stone 2240 pounds = 1 ton	1000 mg = 1 gram 1000 g = 1 kg 1000 kg = 1 tonne	1 kg ≈ 2.2 pounds 1 tonne ≈ 1 ton
Capacity	8 pints = 1 gallon	1000 ml = 1 litre	1 gallon ≈ 4.5 litres

1 The egg of a kiwi-bird has a weight of approximately 0.5 kg.
A fully grown female bird is about 4 times this weight.

   (a) What is the weight of a fully grown female kiwi?

   (b) What is this approximately in pounds?

2 **(a)** An elephant has teeth weighing up to 4.5 kg.
   It has 4 of this size.

   (i) What is their total weight?   **(ii)** Convert this to pounds.

   (b) Each of these teeth is just over 25 cm long.

   (i) How many inches is this?   **(ii)** Is this more or less than a foot?

3 A small whale contains about 400 gallons of pure oil.

   (a) How many pints is this?   **(b)** Change this to litres.

4 Here are the ingredients for shortbread.

5 oz flour   1 oz ground rice
2 oz castor sugar   4 oz butter

Use the conversion graph to write the weights of each ingredient in grams.

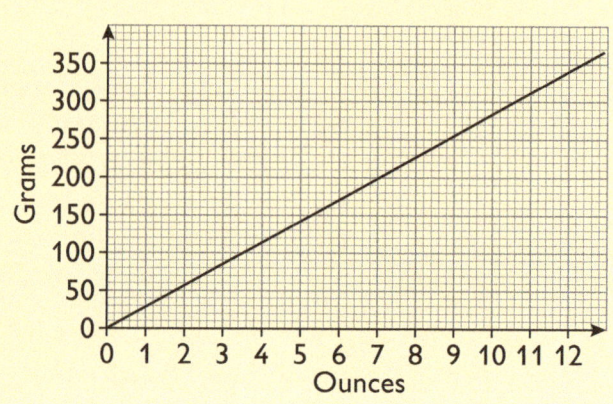

5 The exchange rate for changing euro into Polish zlotys is €1 = 3.2zl.

(a) Draw a conversion graph.
Use your graph to answer the questions below.

> *Your graph should go up to €100.*

(b) (i) Gerry changes €50 into zlotys. How much does he get?

(ii) Ellie buys some wine in Poland. It costs 16zl. How much is this in €?

6 Temperature can be measured in degrees Fahrenheit (°F) or degrees Celsius (°C). This conversion graph is for changing between the two.

(a) Why does this graph not begin at zero?

(b) What temperature in Fahrenheit is equivalent to 0°C?

(c) Normal body temperature is 37°C. What is this in degrees Fahrenheit?

(d) On a very hot summer's day, a news report states that the temperature has reached 90°F. What is this in degrees Celsius?

(e) A comfortable temperature for a room is about 20°C. What is this in degrees Fahrenheit?

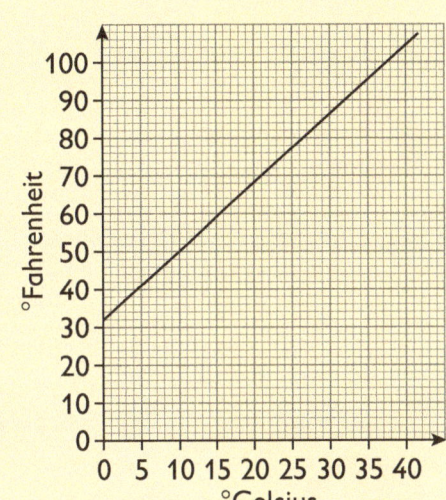

7 Use the conversion graphs in this chapter to convert these amounts.

(a) 4 gallons to litres
(b) 15 pounds to kilograms
(c) 30 miles to kilometres
(d) 250 g to ounces
(e) 60°F to degrees Centigrade
(f) 8 kilograms to pounds

8 Use the approximation 5 miles ≈ 8 km to convert these distances to kilometres.

Birmingham

Birmingham	Brighton	Cambridge	Leicester
160 miles			
100 miles	110 miles		
40 miles	150 miles	70 miles	

9 A Diplodocus was about 86 feet long.

Change 86 feet to (a) inches (b) centimetres (c) metres

10 The world's heaviest man was an American, Jon Minnoch. He weighed 1400 lb. Change this to (a) kilograms (b) tonnes.

12 Shape

Classifying triangles and quadrilaterals

Look at this **decision tree diagram**.

A
B
C
D
E
Start
Choose a triangle
Does it have three equal sides?
No
Yes
Equilateral
Does it contain a right angle?
No
Yes
Does it have two equal sides?
No
Yes
Scalene
Isosceles
Does it have two equal sides?
No
Yes
Scalene right-angled
Isosceles right-angled

? **Use the diagram to classify the orange triangles.**

? **Why are two of the decision boxes ◇ the same?**

? **Can a right-angled triangle be equilateral?**

? **Can an isosceles triangle be equilateral?**

? **Is this the only tree diagram that sorts triangles into these five classes ⬭?**

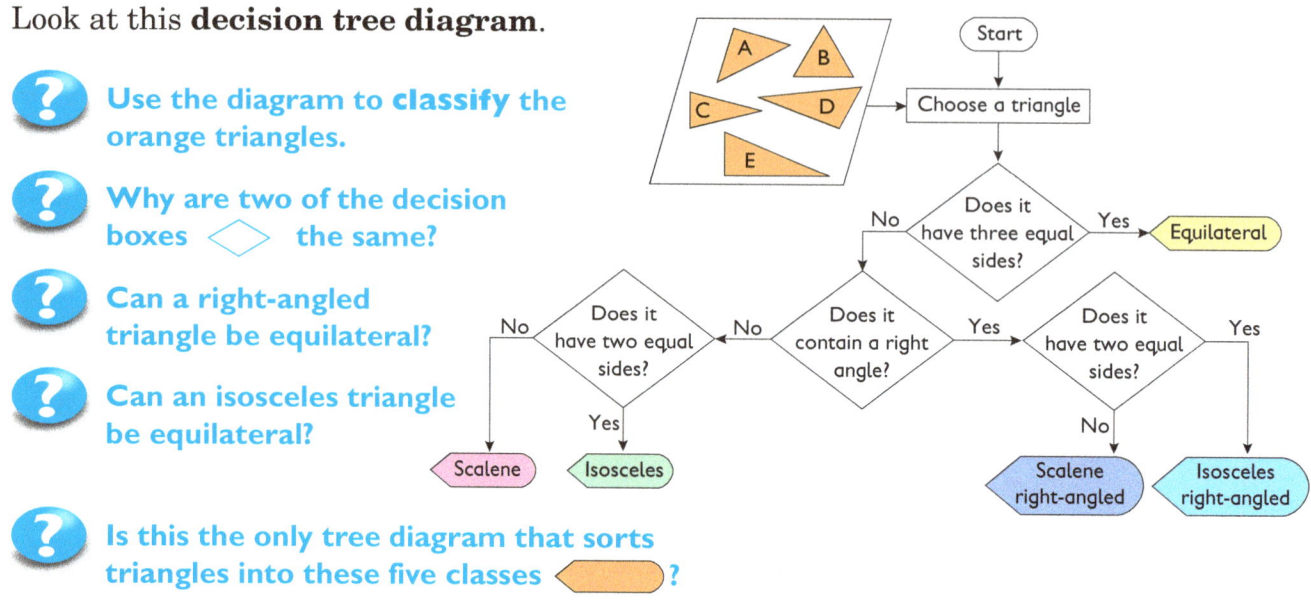

Task

Here is a decision tree diagram to classify quadrilaterals.
Ask your teacher for a copy.

1 Fill in the missing questions.

2 Use the diagram to classify the yellow quadrilaterals.

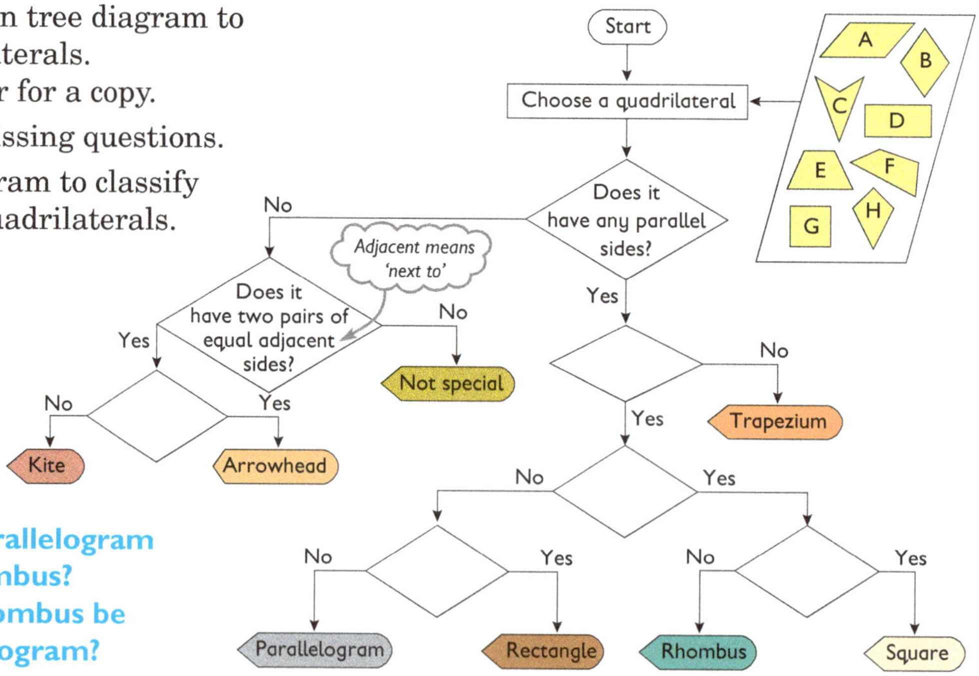

Start
Choose a quadrilateral
A
B
C
D
E
F
G
H
Does it have any parallel sides?
No
Yes
Adjacent means 'next to'
Does it have two pairs of equal adjacent sides?
Yes
No
Not special
No
Yes
Kite
Arrowhead
No
Trapezium
Yes
No
Yes
No
Yes
No
Yes
Parallelogram
Rectangle
Rhombus
Square

? **Can a parallelogram be a rhombus? Can a rhombus be a parallelogram?**

3 Write down five *other* names which could be given to a square.

The decision tree diagrams above sort shapes by length of side, parallel sides and right angles.

Exercise

1 **(a)** Copy this diagram or ask your teacher for a copy.

(b) Write in the missing questions.

(c) Use your diagram to classify the orange triangles.

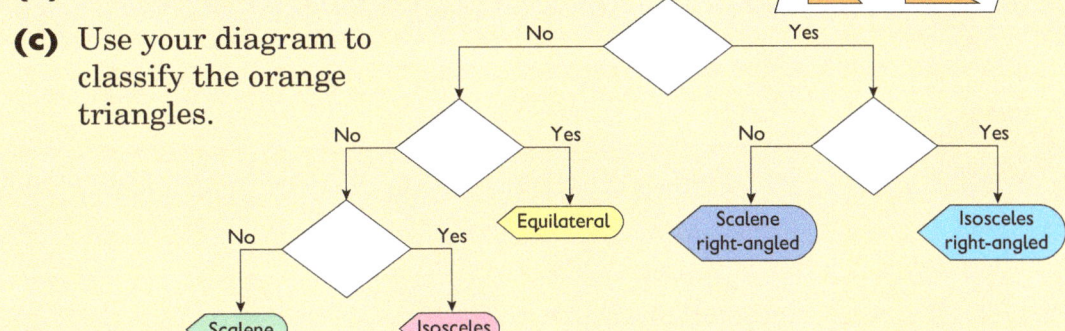

2 Here is *part* of a tree diagram to classify quadrilaterals.
Shape X is a quadrilateral which has some parallel sides.

(a) What does shape X *have* to be?

(b) Can shape X be an arrowhead?

(c) Shape X is also a kite. What other quadrilaterals could X be called?

(d) Copy this diagram and write in the missing questions.

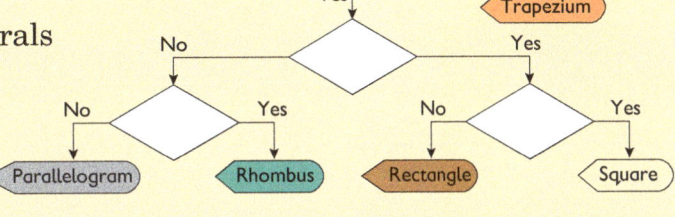

3 Which of the shapes named on page **102** are regular polygons?

Activity Design another decision tree diagram to classify quadrilaterals.
Use these decision boxes:

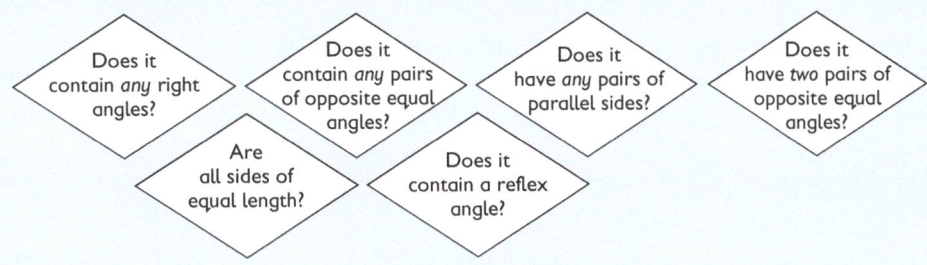

Make sure you include all these outcomes:

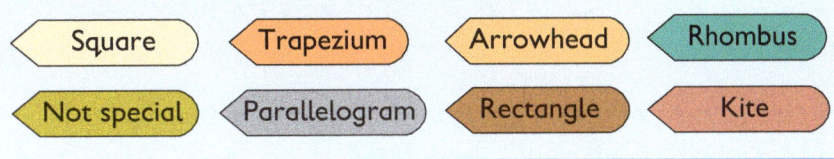

Classifying shapes by symmetry

 How many lines of symmetry does (i) an isosceles (ii) an equilateral (iii) a scalene triangle have?
What about right-angled triangles?

Triangles can be classified by their *symmetry*.

Task

1 Use a pair of compasses and a ruler to construct isosceles triangle AXD.

2 Construct the perpendicular bisector of base AD. This has been started in red.

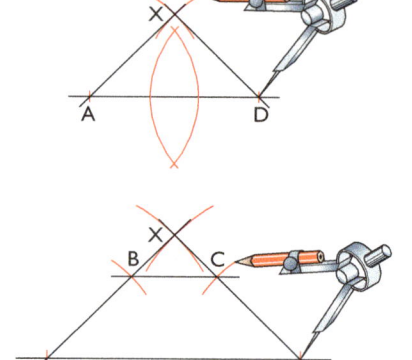

 What can you say about your bisector?
Think of *two* rules for the bisector's locus.

3 Use compasses and a ruler to add line BC to your triangle. AB and DC are equal lengths.

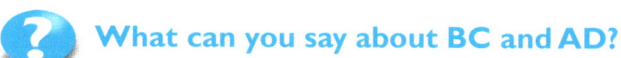 What can you say about BC and AD?

 Quadrilateral ABCD in the Task above is an isosceles trapezium.
How many lines of symmetry does it have?
What can you say about diagonals AC and BD?

 How many lines of symmetry does (i) a rectangle, (ii) a square (iii) a parallelogram have?

Quadrilaterals can be classified by their symmetry.

Task

1 Draw a line and mark points P and R.
2 Construct QS, the perpendicular bisector of PR. This has been started in red.
3 Draw lines to make quadrilateral PQRS.

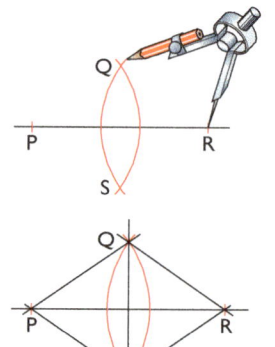

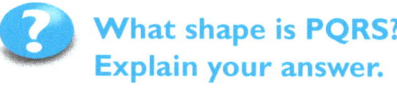 What shape is PQRS?
Explain your answer.

4 PR and QS are the diagonals of PQRS.
You know that QS bisects PR. Explain why PR bisects QS.

 How many lines of symmetry does a rhombus have?
What if the rhombus is also a square?

Exercise

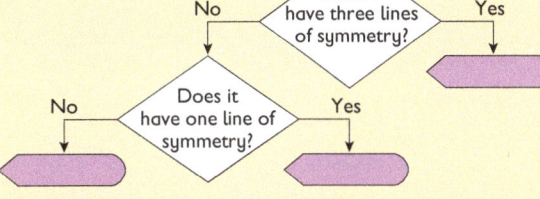

1. **(a)** Copy this decision tree diagram and write in the missing outcomes.

   (b) Is the following statement true or false?

   > All triangles are either scalene or isosceles.

2. **(a)** Follow these steps to construct a kite.

   (b) Draw line XY on your kite. Describe the locus of XY.

   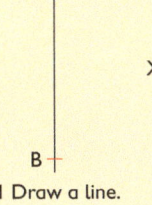 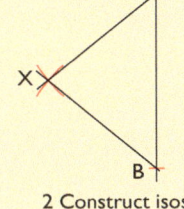 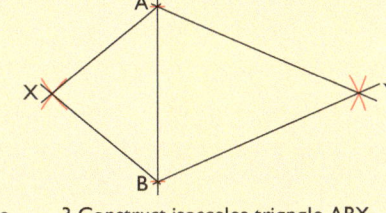

   1 Draw a line. Mark A and B

   2 Construct isosceles triangle ABX.

   3 Construct isosceles triangle ABY. Length AY does not equal length AX.

   (c) Another kite is drawn with length AY equal to AX. What other name can you call this kite?

   (d) Now construct an arrowhead. Write down the steps you take.

3. **(a)** Use squared paper to draw a parallelogram.

   (b) Explain why opposite angles of a parallelogram are equal.

   (c) Draw the diagonals of your parallelogram. Label the intersection X. Measure lengths XA, XB, XC and XD. What do you notice?

   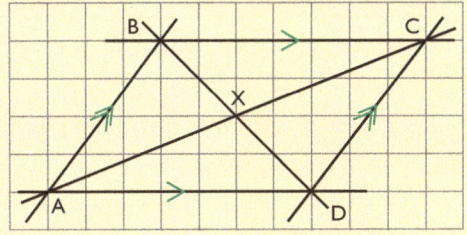

4. Copy and complete this properties table.

Key				
Pa = parallelogram	Sq = square Re = rectangle		Rh = rhombus Ki = kite	

	Quadrilateral				
Property	**Sq**	**Rh**	**Pa**	**Re**	**Ki**
Number of lines of symmetry		2			
Order of rotational symmetry	4				NONE
Are opposite sides parallel?			YES		
Are opposite sides equal length?	YES				
Are *both* pairs of opposite angles equal?					
Is just *one* pair of opposite angles equal?					
Do both diagonals bisect each other?					
Does just one diagonal bisect the other?					
Do the diagonals intersect at right angles?			NO		

Tessellation

Look at this honeycomb.
It is made of cells fitted together.
The cells **tessellate**.

 **What can you say about
(i) the shape (ii) the size of the cells?**

 **There are no gaps between any of the cells.
How far can the pattern of cells continue?**

When *one* shape is repeated to cover a surface
this is called *simple* tessellation.
Here are two ways of tessellating a triangle.

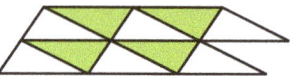 

 How many ways can you tessellate a regular hexagon?

Task

Tessellating quadrilaterals

1 Look at some squared paper.
Do squares tessellate?

2 Using squared paper, draw a rectangle
4 squares by 2.
Find three different ways of tessellating
the rectangle.

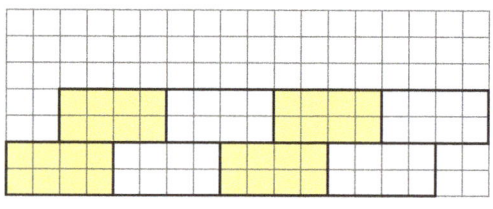

 **Where do you often see patterns of tessellating rectangles?
Can you tessellate a rectangle of *any* length and width? Explain your answer.**

3 Copy this isosceles trapezium onto
squared paper.
Show how the trapezium can be tessellated.

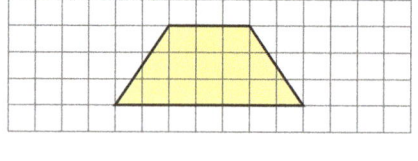

4 Draw some examples to show how you
can tessellate
(a) a trapezium which is *not* isosceles
(b) a parallelogram which is not a rectangle.

 **Can you tessellate *any* trapezium?
Explain your answer.**

5 On a piece of card draw a quadrilateral with
unequal sides between 2 cm and 4 cm.
Cut the quadrilateral out and use it to see if you can tessellate it.

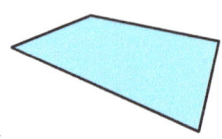

 Explain why *any* quadrilateral will tessellate.

ctivity

This pattern was created by M. C. Escher.
Escher (1898–1972) was a Dutch artist who used
symmetry and transformations to create pictures.

M.C. Escher's "Symmetry drawing E112"
© 2007 The M.C. Escher Company-Holland.
All rights reserved. www.mcescher.com

> **?** **What simple shape has been used for
> this tessellation?**
> **How has the shape been adapted?**

Here you can see a rectangle adapted to make a more interesting shape to tessellate.

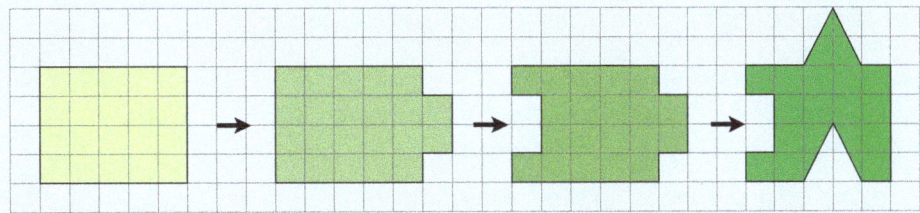

Escher was inspired by tiling patterns in the Alhambra Palace in Spain.
The Alhambra was built by the Moors between 1248 and 1354.
Islamic art is often geometrical.

> **?** **What simple shape has been adapted to create
> this pattern in the Alhambra?**

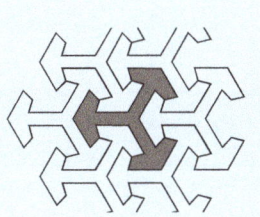

Design your own original shape which will tessellate.
Use squared or isometric paper.
Trace your shape onto card to make a template.
Use your template to 'tile' a piece of paper. Decorate the tiles.

nvestigation

I Look at some isometric paper.
   (a) What type of triangles are on isometric paper?
   (b) What are the interior angles of each triangle?
   (c) Why do these triangles tessellate?

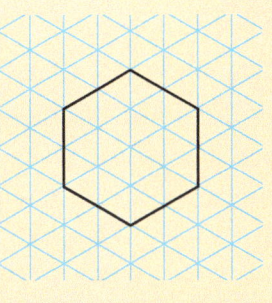

Use isometric paper to draw some tessellating hexagons.
   (d) What are the interior angles of each hexagon?
   (e) Why do regular hexagons tessellate?

The tessellation of a regular polygon is called
a *regular* tessellation.

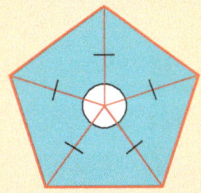

> **?** **Think about the angles in this regular pentagon.
> Explain why a regular pentagon will *not* tessellate.**

2 Copy and complete this table.

> **?** **How many regular tessellations
> are there?**

Regular Polygon	360° ÷ Interior Angle =
Equilateral triangle	$360° \div 60° = 6$
Square	
Regular pentagon	
Regular hexagon	

Solid shapes

Solid shapes are 3-dimensional (3-D).
It is impossible to draw them on paper which is 2-dimensional (2-D).

? **What is a dimension?**

" **Do the right thing!**

Here you can see a cuboid *represented* on isometric paper.

? **Look at the lines on the paper.**
In how many different directions do the lines run?
What does each direction represent?

? **How many edges does a cuboid have?**
How many edges can you see in the diagram?

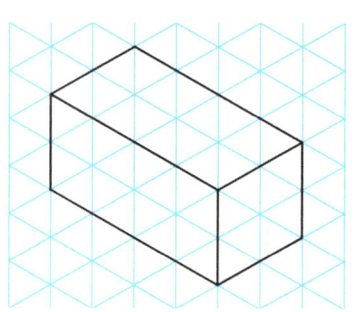

Copy the cuboid onto some isometric paper.
Draw in the *hidden* edges using dashed lines.

Vertical edges of a solid shape are drawn on paper with vertical lines.
Horizontal lines are drawn in any direction *except* vertical.

? **How many horizontal dimensions are there? What are they?**

Task

1 Make a copy of this L-shaped prism onto isometric paper.

Each side of one of the equilateral triangles represents
1 cm on the solid.

2 Calculate **(a)** the area of the L-shaped prism
(b) the volume of the prism
(c) the surface area of the prism.

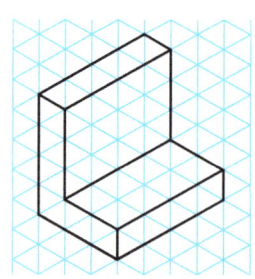

? **Why is this paper called isometric?**

Sometimes a solid shape is represented by
three separate 2-D views.

Task

Here is the front elevation of the C-shaped
prism, drawn on squared paper.

Copy it and add the plan and side elevation.
Label each view.

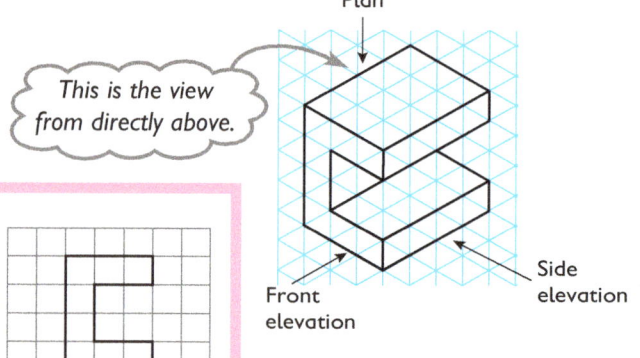

This is the view from directly above.

Plan

Front elevation

Side elevation

Exercise

1 Use isometric paper to make 3-D drawings of prisms with these cross-sections
   (a) a rectangle
   (b) a T-shape
   (c) a right-angled triangle
   (d) a hexagon
   (e) an isosceles triangle.

2 For each isometric drawing in Question 1 used squared paper to construct a plan and two elevations.
   Label each view.

3 Use squared paper to draw the plan and two elevations of these shapes.

(a)    **(b)**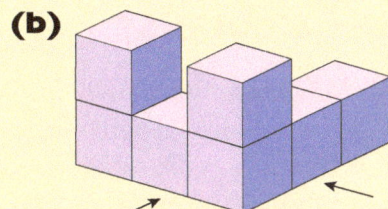

4 The diagram is a plan of a solid shape made from cubes.
   The number in each square is how many cubes are on that base.
   (a) Draw elevations from the two viewpoints shown.
   (b) Make an isometric drawing of the solid.

	3	1
1	2	

5 **(a)** A solid shape is made from 8 cubes joined together.
       Cubes are always joined as shown.
       (i) Find the shape with the smallest surface area.
       (ii) Find the shape with the largest surface area.
          There are several possible answers.
       (iii) Draw the shapes on isometric paper.

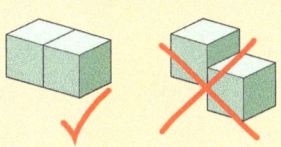

   (b) Repeat this with 12 cubes.

Activity

1 Draw a solid shape using isometric paper.

2 Use squared paper to draw the plan and two elevations of the solid.

3 Swap your plan and elevations with a friend.

Do not look at each other's isometric drawings.

4 Draw the isometric drawing of a friend's solid by looking at their plan and elevations.

5 Now check you have reproduced your friend's isometric drawing.

Try this with another shape.

Finishing off

Now that you have finished this chapter you should:

- be able to classify triangles and quadrilaterals by their geometrical properties
- understand that tessellation is a pattern of shapes repeated to cover a surface with no gaps
- understand different types of tessellation: simple, regular, and non-regular
- understand that where shapes meet to tessellate the angle sum must be 360°
- be able to make 2-D drawings of 3-D solids using isometric paper
- be able to draw and use plans and elevations of solid shapes.

Review exercise

1. Write down the names of the following shapes:
   (a) a quadrilateral with equal sides and equal angles
   (b) a triangle with equal sides
   (c) a triangle with two equal angles
   (d) a polygon with five sides
   (e) a triangle with no equal sides
   (f) a triangle with two equal angles of 45°
   (g) a triangle with two equal angles of 60°.

2. Use squared paper to draw the plan and two elevations of these shapes.

   (a) 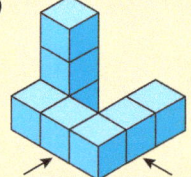　**(b)** 　**(c)**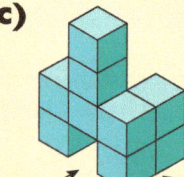

3. Which regular polygons tessellate? Draw a diagram for each tessellation. You will find it helpful to use squared paper or isometric paper.

4. Use squared paper to show how this S shape will tessellate.

5. Use squared paper to show the tessellation of

   (a) an arrowhead　　**(b)** a kite　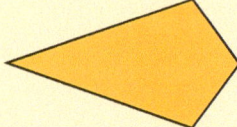

   (c) Show how a kite *and* an arrowhead can tile a surface.

6 Here are the plans and elevations of three solid shapes.
Draw each 3-D shape on isometric paper.

(a)

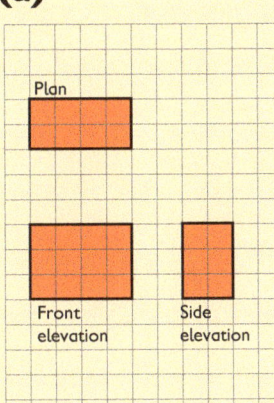

Plan

Front elevation   Side elevation

(b)

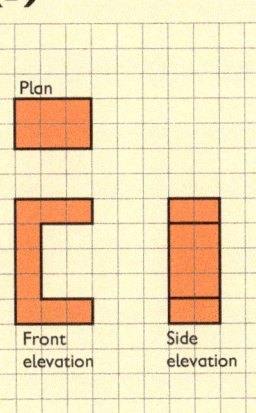

Plan

Front elevation   Side elevation

(c)

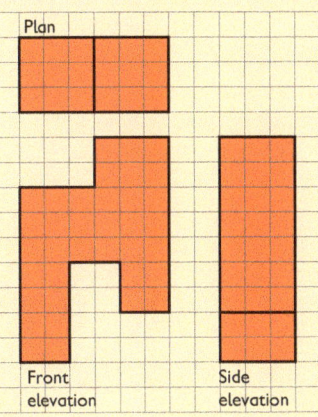

Plan

Front elevation   Side elevation

7 Write down the names of all the quadrilaterals which have

(a) all their sides equal

(b) equal diagonals

(c) two pairs of equal sides

(d) two pairs of equal opposite sides

(e) the diagonals bisecting each other

(f) equal diagonals

(g) the diagonals intersecting at right angles

(h) at least one pair of opposite sides parallel

(i) the diagonals bisecting each other at right angles.

Activity Use *squared* paper to design your own 3-dimensional set of letters or **font**.

Make the front face of each letter 3 or 4 squares wide and 5 squares high.

Here is an example.

Write your name or a message using your font.

Prime factors

The class count to 6.

"one", "fizz", "buzz", "fizz", "five", "fizz, buzz"

 What do they say for 9? For 7?

RULES
THE "FIZZ, BUZZ" NUMBER GAME

Start counting.
If a number is a multiple of 2, replace it with the word "fizz".
Replace all multiples of 3 with the word "buzz".
For any number that is a multiple of 2 and 3 say "fizz, buzz".
Leave all other numbers unchanged.

Johnny replaces 12 with

"fizz, fizz, buzz"

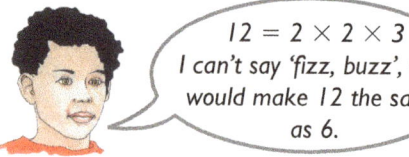

$12 = 2 \times 2 \times 3$
I can't say 'fizz, buzz', that would make 12 the same as 6.

The class decide that every number must sound different. They start the game again.

 What do they say for 4? What about 9 and 8?

Janet notices that they will have to stop at 10.

The class decide to say 'pop' when a number is a multiple of 5.

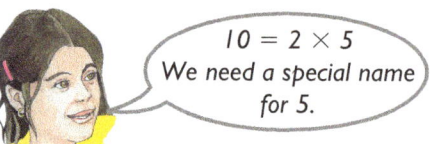

$10 = 2 \times 5$
We need a special name for 5.

 What do they say for 15? How far can they count now?

 Task

Count as far as 21 using the new rules for "Fizz, Buzz" Number Game.
Complete the table to show the numbers that are replaced by names you use.

Name	Number it replaces
Fizz	2
Buzz	3
Pop	5
Bang	–
–	–

Why do you need another new name before you can write 22?

How would you describe the numbers you have written in your table?

Products of primes

You can write any number as the product of its prime number factors.

$$12 = 2 \times 2 \times 3 = 2^2 \times 3$$

Example

To write 72 as a product of primes.

 Write 60 as a product of prime factors.

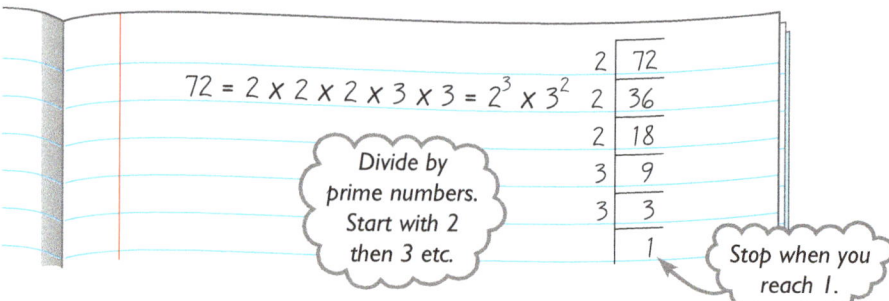

$72 = 2 \times 2 \times 2 \times 3 \times 3 = 2^3 \times 3^2$

2	72
2	36
2	18
3	9
3	3
	1

Divide by prime numbers. Start with 2 then 3 etc.

Stop when you reach 1.

 What has this got to do with the "Fizz, Buzz" Number Game?

Exercise

1 Multiply out the following products

(a) 2×3^2 (b) $2^2 \times 7$ (c) $2 \times 3 \times 5$ (d) $3^2 \times 5$ (e) $2^2 \times 3^2 \times 5$

2 Complete this list of the factors of 36

(a) 1, 2, 3, ..., 36

(b) Which of these factors are prime numbers?

(c) Write 36 as a product of its prime factors.

3 Write the following as products of their prime factors:

(a) 18   (b) 30   (c) 45   (d) 21   (e) 12   (f) 28

(g) 144   (h) 120   (i) 108   (j) 256   (k) 100   (l) 1000

Activity The prime factors of a square number can be written in pairs.

Example $36 = (2 \times 2) \times (3 \times 3)$ and $16 = (2 \times 2) \times (2 \times 2)$
is the square of 2×3 is the square of 2×2

 Explain why $3^2 \times 5^2$ is a square number. What is its square root?
Write down the square of $24 = 2 \times 2 \times 2 \times 3$.
Find the square root of $2^2 \times 3^2 \times 7^2$.

The prime factors of a cube number can be grouped in threes.

Example $216 = (2 \times 2 \times 2) \times (3 \times 3 \times 3)$
This is the cube of 2×3

 Write down the cube of $15 = 3 \times 5$.
What is the cube root of $3^3 \times 7^3$?

$64 = 2 \times 2 \times 2 \times 2 \times 2 \times 2 = 2^6$ is a square number and a cube number.
Show that its factors can be grouped in pairs *and* in threes.
Find two more numbers that are both square numbers and cube numbers.

Investigation When a fraction is written as a decimal, it may go on forever.

These decimals terminate. $\frac{1}{16} = 0.0625$ $\frac{1}{20} = 0.05$

These recurring decimals go on forever. $\frac{1}{3} = 0.333...$ $\frac{1}{7} = 0.142\,857\,142\,8...$

1 Change these fractions to decimals.   (a) $\frac{1}{6}$ (b) $\frac{1}{4}$ (c) $\frac{3}{50}$ (d) $\frac{4}{15}$

2 Write a list of the denominators that produce decimals that do not recur.

3 Write each of these as a product of prime factors.

What can you say about the denominator (bottom line) of any fraction that can be written as a terminating decimal (one that does not recur)?

Lowest common multiple

? **When is the next time that all three buses depart from this stop?**

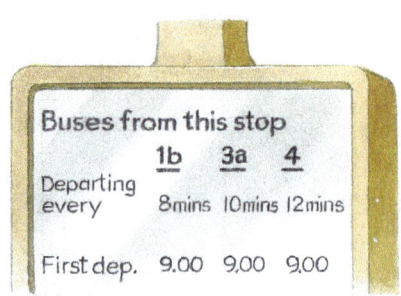

Buses from this stop

	1b	3a	4
Departing every	8mins	10mins	12mins
First dep.	9.00	9.00	9.00

One way to answer this is to find the **lowest common multiple** or **LCM** of 8, 10 and 12. This is the smallest number that two or more numbers will divide into exactly.

Here is each number as a product of its prime factors:

$$8 = 2 \times 2 \times 2 \qquad 10 = 2 \times 5 \qquad 12 = 2 \times 2 \times 3$$

	2	**3**	**5**
8	③	–	–
10	1	–	①
12	2	①	–

The table shows how many of each factor are needed. Choose the largest from each column.

The LCM of 8, 10 and 12 is therefore $2 \times 2 \times 2 \times 3 \times 5 = 120$.

? **How does this tell you the next time at which you could catch any of the buses?**

? **At which times can you catch a *1b* or a *3a* bus but not a number *4*?**

Task

1 Find the LCM of the following sets of numbers:

    (a) 4 and 6   **(b)** 4 and 15   **(c)** 8 and 12

    (d) 12 and 40   **(e)** 60 and 72   **(f)** 5, 6 and 7.

Look carefully at your answers.
Sometimes the LCM of a set of numbers is simply the product of those numbers.

? **Is the LCM of 4 and 15 given by 4 × 15 = 60?**
Is the LCM of 8 and 12 given by 8 × 12 = 96?

2 Explain your answers.

Fractions and the LCM

The **lowest common denominator** of two or more fractions is the LCM of all the denominators.

? **Find the LCM of 16, 12 and 15.**
Write the fractions $\frac{3}{16}, \frac{5}{12}$ and $\frac{4}{15}$ with a common denominator and place them in ascending order.

Exercise

1 Find the LCM of the following sets of numbers:

(a) 8 and 10   (b) 3, 5 and 6   (c) 30 and 40   (d) 16 and 20

(e) 12 and 21   (f) 120 and 180   (g) 120, 180 and 200

2 Use your answers from question 1 to help you work out the following

(a) $\frac{1}{8} + \frac{1}{10}$ (b) $\frac{1}{3} + \frac{1}{5} + \frac{1}{6}$ (c) $\frac{7}{30} - \frac{3}{40}$ (d) $\frac{5}{16} - \frac{3}{20}$

3 Alan and David are walking side by side. They are in step.
Alan's stride is 80 cm long. David's stride is 90 cm long.
How far do they walk before they are next in step?

4 The dials below are set at 0.

After the left dial has been turned once they will look like this.

(a) Draw a diagram to show what they will look like after two complete turns of the left dial.

(b) How many complete turns of the left dial are needed before the first two dials are both set to 0?

(c) How many complete turns of the left dial are needed before all three dials are again set to 0?

Activity In a motor race, the faster car completes the circuit in 5 minutes and the slower car takes 7 minutes.

1 (a) Find the LCM of 5 and 7.

(b) (i) How long is it before the faster car overtakes the slower car as they pass through the starting position?

(ii) How many laps has each car completed?

(c) (i) Does the faster car overtake the slower car anywhere else on the circuit?

(ii) If so, where does this happen and after how long?

Repeat (a) to (c) above when the times of the two cars are:

2 6 minutes and 9 minutes   **3** 5 minutes and 8 minutes.

Highest common factor

The class play the common factor game.

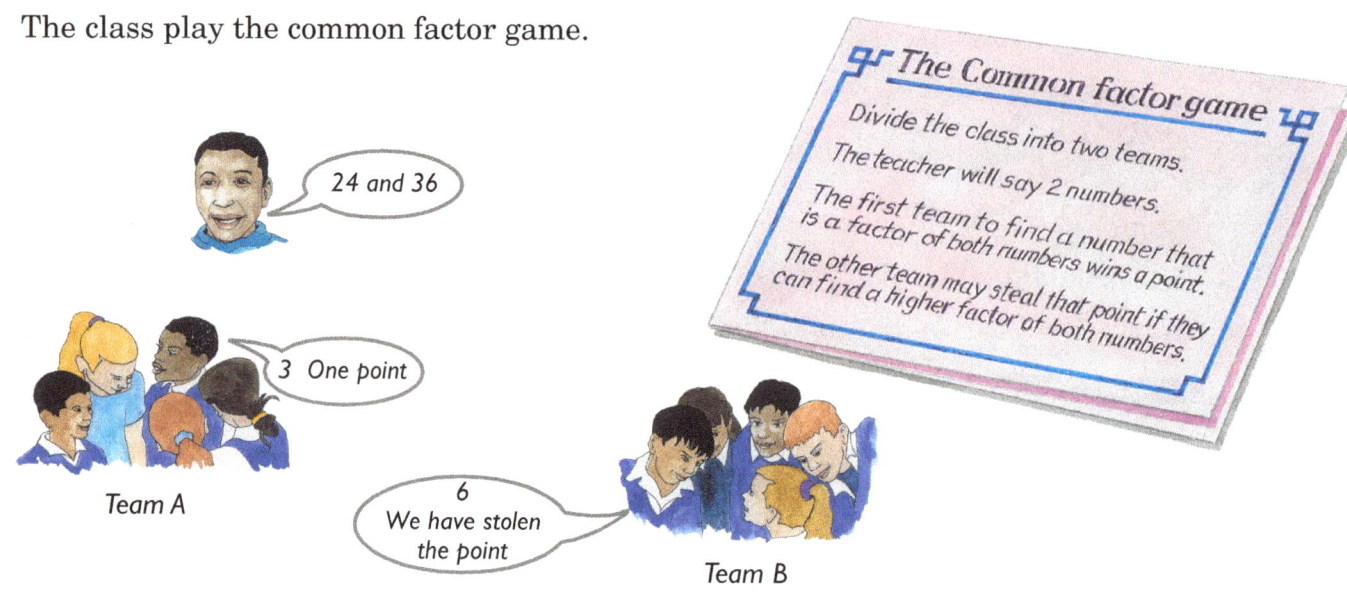

Team A

Team B

 Is it possible for team A to steal the point back?

 What is the winning number for 30 and 45? What about 4 and 15?

Task

Play the Common Factor Game.
Make a careful note of the winning number for each pair of numbers.
The winning number is called the **highest common factor** or **HCF** of the numbers.
This is the *largest* number that is a factor of each of the numbers.

 When is the HCF of two numbers 1?

The HCF can be found when both numbers are written as a product of their prime factors.

$$24 = 2 \times 2 \times 2 \times 3 \qquad 36 = 2 \times 2 \times 3 \times 3$$

	2	3	5
24	3	①	–
36	②	3	–

Choose the smallest number from each column this time.

The HCF of 24 and 36 is $2 \times 2 \times 3 = 12$.

 Which is bigger, the HCF of two numbers or the LCM?
Work this out for   (i) 18 and 24   (ii) 72 and 180.

 Are the HCF and LCM ever the same?

Exercise

1 Find the HCF of the following numbers.

(a) 12 and 8   (b) 12 and 4   (c) 7 and 8   (d) 30 and 45

(e) 5, 10 and 20   (f) 78 and 18   (g) 140 and 210   (h) 52 and 65

2 The map shows the route of a charity walk.

Marshals are to be placed at equally spaced intervals along the route, including at the start and finish. There must be a marshal at the lunch stop.

(a) Find the HCF of 30 and 18.

(b) What is the least number of marshals that are needed?

3 The area of garden shown is to be covered with square paving stones.

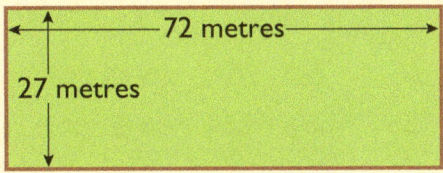

(a) Find all the common factors of 27 and 72.

(b) What is the largest size for each stone?

4 Sanjeev makes rosewood jewellery.
Pieces of wood are joined together to make bracelets, necklaces and anklets.
All the pieces of wood are the same length.
Look at the poster on the right.

(a) Write down the common factors of 18, 48, 54 and 24.

(b) What are possible lengths for the pieces of wood?

(c) What length do you think is best?

5 Class 8c knit squares as large as possible to join together into blankets. The blankets are made in 2 sizes.

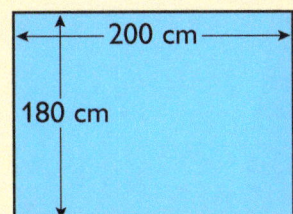
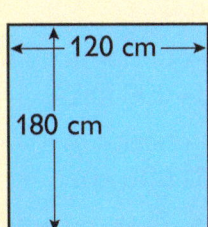

(a) Find the HCF of 200, 180 and 120.

(b) How large should they make the squares?

Finishing off

Now that you have finished this chapter you should be able to:

- write any number as the product of its prime factors
- find the lowest common multiple of a set of numbers
- find the highest common factor of a set of numbers
- use the LCM and HCF when solving problems.

Review exercise

1 Multiply out the following:

(a) $3^2 \times 7$ **(b)** $2^3 \times 5$ **(c)** $2^3 \times 7^2$ **(d)** $2 \times 3 \times 7 \times 11$ **(e)** $7^2 \times 11$

2 Write each of these numbers as a product of its prime factors:

(a) 8 **(b)** 15 **(c)** 20 **(d)** 50 **(e)** 70 **(f)** 240

(g) 312 **(h)** 729 **(i)** 52 **(j)** 224 **(k)** 11 240

3 Find the LCM of the following sets of numbers:

(a) 2 and 3 **(b)** 2 and 5 **(c)** 2 and 4 **(d)** 6 and 10

(e) 12 and 15 **(f)** 20 and 30 **(g)** 45 and 120 **(h)** 15, 20 and 12

(i) 12 and 64 **(j)** 7, 8 and 9 **(k)** 6, 8 and 10 **(l)** 6, 9 and 12

4 Find the HCF of the following sets of numbers:

(a) 2 and 4 **(b)** 3 and 9 **(c)** 6 and 10 **(d)** 8 and 18

(e) 15 and 18 **(f)** 6, 9 and 15 **(g)** 52 and 65 **(h)** 7, 8 and 9

(i) 42 and 70 **(j)** 66 and 121 **(k)** 72, 36 and 18 **(l)** 63 and 198

5 Anne's father services his car
every 6000 km.
Here are some of the checks
he carries out.

(a) Which checks are needed
after 24 000 km?

(b) Which checks are needed
after 54 000 km?

(c) A total overhaul requires
all the checks.
How often is this needed?

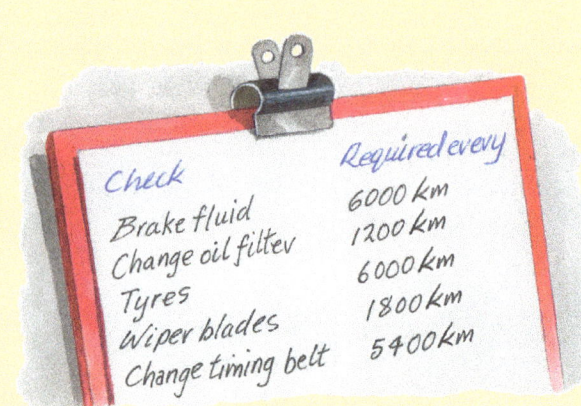

Check — Required every

Brake fluid — 6000 km

Change oil filter — 1200 km

Tyres — 6000 km

Wiper blades — 1800 km

Change timing belt — 5400 km

6 Below is a map of Puffin Island.

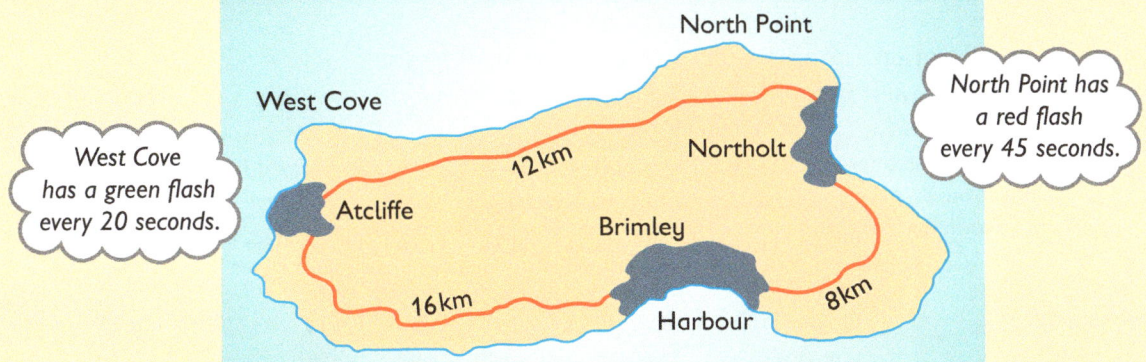

North Point

West Cove

12 km

Northolt

North Point has a red flash every 45 seconds.

West Cove has a green flash every 20 seconds.

Atcliffe

Brimley

16 km

8 km

Harbour

(a) Both lighthouses emit a flash at 6.00 pm. When do they next both emit a flash?

(b) There are telephones in each of the towns.
New telephone boxes are to be placed at equal distances around the coast road.
Suggest how far apart these should be.

(c) The island bus arrives at the harbour every 12 minutes.
Ferries depart every 20 minutes for the mainland and every 45 minutes for the neighbouring island of Galleon.

The 9.00 am bus connects with both ferries.
At what time does this next happen?

(d) What is the longest time that you will have to wait for the mainland ferry if you arrive by bus?

Activity Cut out 6 equal cardboard circles. Label them 3, 4, 6, 9, 12 and 18.
Cut evenly spaced notches from the edge of each circle.

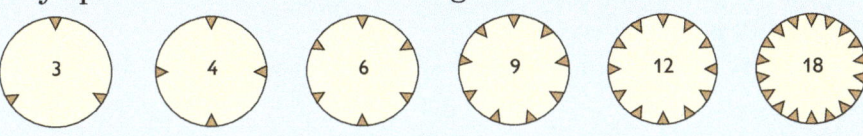

1 Place circle 6 on top of circle 9.
Rotate the circles so that one notch on circle 6 is lined up with one notch on circle 9.
How many more notches are lined up? Explain.

2 Repeat the above for circle 12 and circle 18.
How many notches match when circles 4, 6 and 12 are placed on top of each other?

3 The circles can also be labelled as:

To each circle	3	4	6	9	12	18
Add the labels	120	90	60	40	30	20

Explain the new labels.
Experiment with them to find the LCM of different pairs of numbers.

14 Doing a survey

Gino Marconi is opening an ice-cream parlour in Avonford.
He wants to find out if it will be successful. He asks some people on the High Street a question.

? **Do these answers help him to decide** **(a) how often people will come**
(b) what flavours to sell **(c) how much to charge** **(d) which hours to open?**

Task

Write down 4 questions which would give Gino more useful information.

He needs to collect *useful* data. The best way to do this is to use a questionnaire or a survey.

Designing a questionnaire

Think about the following points when you are designing a questionnaire.

1 Where possible give a choice of responses.
Bad How often are you late for school?

> Too many possible answers.

Better How many days were you late for school last week?

☐ ☐ ☐ ☐ ☐ ☐
0 1 2 3 4 5

2 Allow for all possible responses.
Bad How many miles do you cycle per week?

☐ ☐ ☐ ☐
0–10 11–20 21–30 31–40

> What about people who do more than 40?

Better How many miles do you cycle per week?

☐ ☐ ☐ ☐ ☐
0–10 11–20 21–30 31–40 more than 40

3 It must be clear which box to tick.
Bad How old are you?

☐ ☐ ☐ ☐ ☐
0–10 10–20 20–40 40–60 over 60

> A 10-year-old has two possible boxes.

Better How old are you?

☐ ☐ ☐ ☐ ☐
0–9 10–19 20–39 40–59 more than 60

> Sometimes you may expect people to tick two or more boxes. In that case you must say so.

4 Responses should be balanced.
Bad What do you think of the pop group Take This?

☐ ☐ ☐ ☐
excellent very good quite good average

Better What do you think of the pop group Take This?

☐ ☐ ☐ ☐ ☐
very good good average poor very poor

> There should be the same number of negative and positive responses.

Exercise

1 Design a question for each of the following sets of responses.

(a)
☐ always ☐ sometimes ☐ never

(b)
☐ 0 ☐ 1 ☐ 2 ☐ 3 ☐ 4 ☐ more than 4

(c)
☐ strongly agree ☐ agree ☐ neither agree or disagree ☐ disagree ☐ strongly disagree

(d)
☐ less than €1 ☐ between €1 and €4.99 ☐ between €5 and €9.99 ☐ €10 or more

(e)
☐ less than 10 mins ☐ at least 10 mins but less than 20 mins ☐ at least 20 mins but less than an hour ☐ at least an hour

2 In each of the following cases

(i) state what is wrong with the question
(ii) design a better question.

(a) What do you think of the computer game Crypt Stormer?
☐ excellent ☐ very good ☐ quite good ☐ average

(b) How many packets of crips do you eat in a week?
☐ 0–4 ☐ 5–9 ☐ 10–14 ☐ 15–19

(c) How many days were you off sick last year?
☐ 0 ☐ 1 ☐ 2 ☐ more than 2

3 Write down a set of responses for each of the following questions.

(a) How much time do you spend watching TV each week?

(b) When was the last time you visited a health centre?

(c) How much would you be prepared to pay for cable television each month?

Activity Sarah is thinking of starting a lunchtime club.
Design a short questionnaire to help her decide

(a) what kind of club to start

(b) what day to hold it on

(c) whether enough people would attend.

Biased and misleading questions

Gino continues his survey.

1 Do you like to keep cool and refreshed on a hot day?

☐ Yes   ☐ No

2 Is ice-cream cool and refreshing?

☐ Yes   ☐ No

3 Would you visit the ice-cream parlour?

☐ Yes   ☐ No

Gino's questions are **biased**.
Most people will answer 'Yes' to Questions 1 and 2.
This **leads** them to answer Question 3 with a 'Yes' as well.

**Task**

Make up two questions which could be placed before Question 3 which might make people answer *No*.

Avoid questions which make your own feelings obvious.

? **What is wrong with these two questions?**

(a) I think the pop group Blokezone are brillant. What do you think of them?

☐ very good   ☐ good   ☐ average   ☐ below average   ☐ poor   *Tick one box*

(b) The Star 1800 car is rubbish because

☐ it does not go fast enough   ☐ it is uncomfortable   ☐ it does not have a CD player

The results of the survey can be biased by
(a) asking people who are not typical   **(b)** having someone special ask the questions.

? **What is wrong with this situation?**

Do you own a motorbike?

☐ Yes   ☐ No

Task

Design a short unbiased questionnaire to help Gino find out the information he wants.

It is not easy to design a good questionnaire.
Always try one out on a small group of people first. This is called a **pilot survey**.

? **What action do you take after your pilot survey?**

Exercise

1 All these questions are biased. Say why and rewrite them.

(a) Sausages taste great don't they?

☐ ☐
Yes No

(b) We think fox-hunting should be banned. Do you?

☐ ☐ ☐ ☐ ☐
strongly agree   agree   neither agree   disagree   strongly disagree
nor disagree

(c) Why do you think computer games should be banned?

☐ ☐ ☐
Students play them instead   They are   Sitting for a long time at a
of doing homework   too violent   computer is bad for your body

2 You want to find people's views on the following subjects.
For each case, write down **(i)** a group of people who are not suitable to ask
(ii) a person who should not ask the question.

(a) Shakespeare's plays

(b) The standard of teaching at St Mugwump's College

(c) Professional boxing

3 St John's College is considering opening a tuck shop.
They want to find out ● what to sell ● when to open
● whether it will make a profit.

This questionnaire has been designed for students to answer.

It has been badly written.

---

1 Would you like a Tuck Shop?.................

2 What would you like it to sell?

☐ ☐
chocolate   crisps

     St John's
     College

3 Do you like fizzy drinks?

☐ ☐ ☐ ☐
very much   lots   yes   no

4 How much do you think you would spend per week in the Tuck Shop?

☐ ☐ ☐
less than 50 cent   between 50 cent and €1   between €1 and €2

5 The Tuck Shop should be open before school.

☐ ☐ ☐ ☐
strongly disagree   disagree   agree   strongly agree

6 The Tuck Shop should be open at break.

☐ ☐ ☐ ☐
strongly agree   agree   disagree   strongly disagree

7 You would like the Tuck Shop to be open at lunchtime, wouldn't you?

☐ ☐
yes   no

8 Do you often visit the local Newsagents?

☐ ☐
yes   no

(a) Write down as many errors as you can find.

(b) Rewrite the questionnaire.

Finishing off

Now that you have finished this chapter you should be able to:

● carry out a survey of your own, following the points given below.

Carry out your own survey

1 Decide what the survey is going to be about.

4 Draw up your final questionnaire.

5 Carry out the survey.

We must give it to everyone.

We must try to get all the surveys back too.

It can't just be our friends.

6 Collect the results.

Everyone's got to help.

7 Write your report. Use the following sections.

Aim	What were you trying to find out from your questions?

Method	Who did you ask? How did you choose them? Who asked them?

Questionnaire	Include a copy of your questionnaire.

Results	Give the results of each question as a frequency table and a data display.

Conclusions	What are the answers to your questions? Could you have done anything better?

Ratio and proportion

Mark is cooking Tuna and Cheddar Stuffed Peppers.
His recipe serves **4 people**.

4 red peppers
100 g bread crumbs
4 tomatoes
30 ml pickled chillies
50 g mild cheddar cheese
400 g tinned tuna

For one person the quantity of breadcrumbs is $\frac{100}{4} = 25$ g.

> **These quantities must all be kept in proportion.**
>
> When the amount of one ingredient is divided by 4, all the others must be too. Otherwise the flavour will be altered.

 What are the quantities of the other ingredients for one person.

 **Task**

Mark also cooks a starter (Consommé Julienne) and a dessert (Peach Melba).
Work out the quantities for one person for each recipe.

Consommé Julienne (for 8 people)

3.2 litres of fat-free stock
4 leeks
4 celery sticks
4 shallots
4 egg whites

Peach Melba (for 6 people)

6 meringue nests
336 g fresh raspberries
36 g caster sugar
600 g peach slices

Mark is actually cooking the meal for 7 people.
What quantities of food does he need for each of the 3 courses?

 How did finding the amounts for one person help you find the amounts for 7 people?

 Why is this method called the unitary method?

Exercise

1 Mrs Jamieson is making up party bags for Fiona's party.
There will be 30 people.

  (a) A six-pack of chocolate wheels costs 84 cent.
   (i) How much does one chocolate wheel cost?
   (ii) How much do 30 cost?

  (b) A five-pack of nut bars costs €1.20.
   (i) How much is one nut bar?
   (ii) How much are 30 bars?

  (c) A pack containing 10 mini-rolls costs €1.80.
   (i) How much is one mini-roll?
   (ii) How much for 30 mini-rolls?

  (d) Find a quicker way to work out these answers.

2 A pack of three video tapes costs €8.49.
  (a) How much does one tape cost?
  (b) How much do five tapes cost?

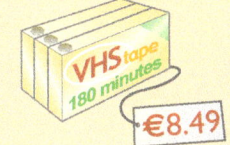

3 Roses are priced at €4.80 for a bunch of eight.
What is the cost of 24 roses?

4 Christmas cards are wrapped in packs of ten.
Each pack costs €4.50.
The shopkeeper sells the cards separately.
What is the cost of 14 cards?

5 One mole of carbon (C) combines with one mole of oxygen (O_2) to make
one mole of carbon dioxide (CO_2).

   $C + O_2 \rightarrow CO_2$

How many moles of oxygen combine with
  (a) 3 moles of carbon   **(b)** 7 moles of carbon?

6 Two moles of hydrogen (H_2) combine with one mole of oxygen (O_2) to make
two moles of water (H_2O).

   $2H_2 + O_2 \rightarrow 2H_2O$

How many moles of hydrogen combine with
  (a) **(i)** 4 moles of oxygen   **(ii)** 6 moles of oxygen?
  (b) How many moles of H_2O are produced in each case?

7 Look at this recipe
for 30 truffles.

  (a) How much of each ingredient
   is in each truffle?
  (b) How much is needed to make
   (i) 20 truffles
   (ii) 50 truffles?

Ratios

Tony is a paralympic champion.
He has a ramp up to the door of his house.
There are supports every 20 cm to make
the ramp stronger.

25 cm

C ←———— 1 m ————→ B

Task

1   Make an accurate scale drawing of the ramp.
    Use a scale of 1 cm to represent 5 cm.

2   Mark the points B_1, B_2, B_3, … every 20 cm
    from C.

3   With your protractor draw an angle of 90° at B_1
    and mark A_1.

    Mark A_1B_1 and write down the ratio $\dfrac{A_1B_1}{CB_1}$.

4   Now mark A_2.

    Measure A_2B_2 and write down the ratio $\dfrac{A_2B_2}{CB_2}$.

5   Continue to the end of the ramp.

    Change these fractions to decimals.

6   What do you notice?

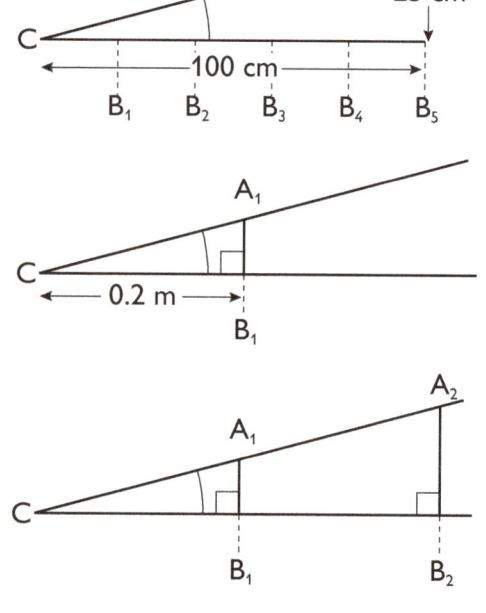

You must always use the same units in ratios. The height of Tony's ramp is
25 centimetres and the distance along the ground is 1 metre.

The ratio is *not* 25 : 1. It is 25 : 100 because 1 metre is 100 cm.

Ratios can be simplified.
You cancel them down like fractions.
To simplify 12 : 8, divide by 4. You get 3 : 2.

?   **How do you simplify 20 : 5?**

?   **These ratios have been simplified.**
       How has it been done?
       (a) 6 : 18 : 24 = 1 : 3 : 4   **(b) $3\frac{1}{2} : 5\frac{1}{2} = 7 : 11$.**

Exercise

1 In a box there are 20 pink, 40 peach and 30 white tissues.
   (a) Write down the ratio, pink to peach to white tissues.
   (b) Simplify your answer.

2 A box contains 6 plain chocolates and 10 milk chocolates.
   (a) Write down the ratio of plain chocolates to milk chocolates.
   (b) Simplify your answer.
   (c) 42 plain chocolates are put into boxes.
       How many milk chocolates are needed to keep the ratio the same?
   (d) 40 milk chocolates are put into boxes.
       How many plain chocolates are needed to keep the ratio the same?

3 A florist puts 3 yellow and 5 white roses into each bunch of flowers.
   (a) Write down the ratio of yellow to white roses.

   One afternoon she uses 18 yellow roses.
   (b) How many white roses does she need to keep the ratio of yellow to
       white the same?

4 Simplify the following ratios.
   Some ratios have different units.
   You must change to the same units first.

   (a) $36:24$ **(b)** $30:48$ **(c)** €16 : €12
   (d) 40 hours : 16 hours   **(e)** $9:15:18$ **(f)** $24:36:54$
   (g) €1.25 : 85 cent (work in cent)   **(h)** 4 m : 360 cm (change to cm)
   (i) 2 hours : 150 minutes   **(j)** 32 mm : 1.6 cm
   (k) $4\frac{1}{2}:2\frac{1}{2}$ **(l)** $2\frac{1}{2}:7\frac{1}{2}$

5 Methane consists of 1 mole of carbon and 2 moles of hydrogen.
   (a) Write down the ratio, moles of carbon to moles of hydrogen.
   (b) How many moles of carbon are needed to combine with 14 moles of
       hydrogen?
   (c) How many moles of hydrogen are needed to combine with 10 moles
       of carbon?

6 In chemistry, two moles of magnesium combine with one mole of oxygen to
   form two moles of magnesium oxide.
   (a) What is the ratio, moles of magnesium : moles of oxygen?
   (b) How many moles of oxygen combine with 6 moles of magnesium?
   (c) How many moles of magnesium combine with 2 moles of oxygen?
   (d) Some magnesium oxide contains 4 billion moles of magnesium.
       How many moles of oxygen does it contain?

Sharing in a given ratio

Tony, Joe and Sophie buy a set of 60 old football programmes between them.
The set costs €180.

Tony pays €30. Joe pays €60. Sophie pays €90.

That is 20 programmes each.

Not fair!

? **Why does Sophie say 'Not fair!'?**

You can use ratio to decide how to share things fairly.
Look at the ratio of money paid by the three friends who bought the football programmes.

Tony : Joe : Sophie = 30 : 60 : 90
                            = 1 : 2 : 3

Altogether 1 + 2 + 3 = 6 shares.

6 shares = 60 programmes

So 1 share = $\frac{60}{6}$ = 10 programmes

Tony gets $1 \times 10 = 10$, Joe gets $2 \times 10 = 20$ and Sophie gets $3 \times 10 = 30$ programmes.

Task

Mrs McTaggart, Mr O'Flynn and Ms Jones buy a sports shop together.
Mrs McTaggart spends €500 000, Mr O'Flynn €600 000 and Ms Jones €900 000.

1 In the first year they make €60 000 profit. How much should each of them get?

2 Ms Jones's share of the second year's profit is €36 000.
How much do Mrs McTaggart and Mr O'Flynn get?

Exercise

1 Divide €24 between two people in the ratio 5 : 3.
How much does each person get?

2 Divide 96 cm into three parts in the ratio 4 : 1 : 3.
How long is each part?

3 A school fete raises €2870.
It is shared between the school and a charity in the ratio 4 : 3.
How much does each receive?

4 An electricity bill is €288.
The ratio of the electricity used for heating and lighting is 7 : 2.
How much is spent on heating and how much on lighting?

5 Pastry is made with twice as much flour as fat by weight.

(a) What is the ratio of the weight of the flour to the weight of the fat?

(b) The fat and the flour together weigh 450 g.
How much flour is there? How much fat?

6 90 boys were asked for their favourite game.
Their answers are shown in the pie chart.

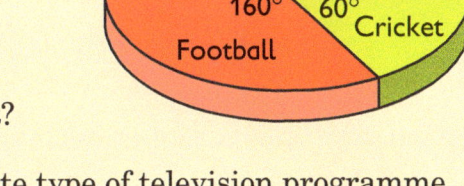

(a) (i) Write down the ratio of the
angles, rugby : football : cricket.

(ii) Simplify this ratio.

(b) How many boys like each game best?

7 In a survey 180 people give their favourite type of television programme.

The results were

Sport — 54
News — 30
Soap Operas — 60
Drama — 20
Others — 16

(a) (i) Write down the ratio of
Sport : News : Soap Operas : Drama : Others.

(ii) Simplify this ratio.

(b) Divide 360° in this ratio.

(c) Draw a pie chart to illustrate these data.

Finishing off

Now that you have finished this chapter you should be able to:

- write down a ratio
- simplify a ratio
- use the unitary method and direct proportion
- divide a quantity in a given ratio.

Review exercise

1 Simplify the following ratios:

(a) 8 : 2   **(b)** 5 : 10   **(c)** 10 : 5   **(d)** 12 : 20

(e) 35 : 25   **(f)** 16 : 24   **(g)** 50 : 45 : 30   **(h)** 24 : 16 : 32

2 For the following

     (i) change them to the same units
     (ii) simplify them.

(a) €1 : 50 cent   **(b)** 10 km : 600 m

(c) 8 hours : 1 day   **(d)** 2 days : 1 week

(e) 2 cm : 5 mm   **(f)** 20 cm : 1 m

(g) 3 minutes : 80 seconds   **(h)** 2 litres : 50 ml

3 Divide 48 sweets between Toni and Kirsty in the ratio 7 : 9.

4 Divide €48 000 profit between Mr Shaw, Mrs Taylor and Miss Weekes in the ratio of 9 : 4 : 3.

5 A school raises €336 for charity.
The money is divided between the NSPCC and the Save the Children Fund in the ratio 3 : 5.
How much does each charity receive?

6 The scale of a model aeroplane is 1 to 200.
The length of the model is 12 cm.
What is the length of the aeroplane?

12 cm

7 A scale on a map is 1 : 10 000.

(a) Find the actual distance, in metres, which is represented by 1 cm.

(b) Find the actual distance represented by 4 cm on the map.

8 A map scale is given as 1 : 1 000 000.
On the map, Bristol and Gloucester are 5.4 cm apart.
What is the real distance between Bristol and Gloucester?

9 A man gives a present of €6000 to be shared between his two sons in the ratio of their ages. They are 13 and 17 years old.
How much does each receive?

10 **(a)** Here are the ingredients in a recipe for Soft Fruit in Summer Sauce.

> **Soft Fruit in Summer Sauce**
> **225 g redcurrants**
> **450 g raspberries**
> **125 g caster sugar**

    (i) Write down the ratio of redcurrants : raspberries.

    (ii) Simplify this ratio.

    (iii) Write down the ratio of redcurrants : raspberries : caster sugar.

    (iv) Simplify this ratio.

(b) A different recipe gives the ingredients in imperial units.

> 8 ounces redcurrants
> 1 pound (= 16 ounces) raspberries
> 4 ounces caster sugar

This recipe serves 8 people.

    (i) Are the ratios the same in the two recipes?

    (ii) Work out the ingredients for each person.

    (iii) Work out how much of each ingredient is needed for 6 people.

11 240 teenagers were asked to vote for their favourite pop group.

The votes were

Roses	60
Step On	30
East 19	70
Pepper Boys	80

(a) **(i)** Write down the ratio, Roses : Step On : East 19 : Pepper Boys.

    (ii) Simplify this ratio.

(b) Divide 360° in this ratio.

(c) Draw a pie chart to illustrate the above data.

Sue replaces the glass in one end of her greenhouse.
She works out the total area of glass needed.

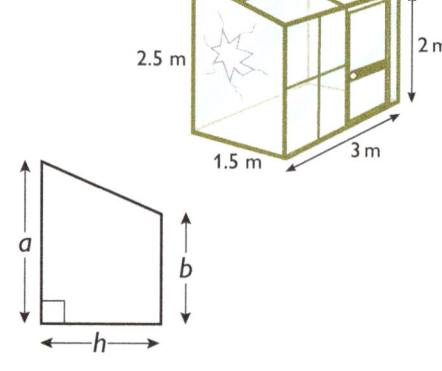

 What shape is the side with the broken window?

She uses the **formula**:

> Area of a trapezium $= \frac{1}{2}(a + b)h$

 What are the values of a, b and h?

Sue **substitutes** the values of a, b and h into the formula.
She works out the area:

Area of trapezium $= \frac{1}{2}(a + b)h$

$= \frac{1}{2} \times (a + b) \times h$

> Replace a, b and h with 2.5, 2 and 1.5.

$a = 2.5$, $b = 2$ and $h = 1.5$

So area $= \frac{1}{2} \times (2.5 + 2) \times 1.5$

$= \frac{1}{2} \times 4.5 \times 1.5$

> Work out the brackets first.

$= 3.375$

So Sue needs 3.375 m² of glass.

 How can you check that the formula works?

Task

The area of this trapezium is 36 cm².

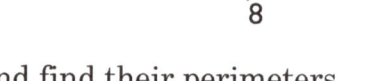

Not to scale

1 What is the value of h?
2 Draw the trapezium accurately.
3 Measure the sloping side and so find the perimeter of the trapezium.
4 Draw two different trapezia, both with area 36 cm², and find their perimeters.
5 What is the smallest possible perimeter for a trapezium with area 36 cm²? Only use whole number values for a, b and h.

Sue wants to make sure that the temperature in her greenhouse is 77°F.
Her thermometer only has a Celsius scale.

She uses the formula $C = \dfrac{5(F - 32)}{9}$ to convert from Fahrenheit to Celsius.

 What do C and F stand for?   What is the value of C when $F = 77$?

Exercise

1 Find the value of $7a + 4b - 3c$ when

(a) $a = 2$, $b = 5$ and $c = 3$ (b) $a = 6$, $b = -7$ and $c = 2$

(c) $a = 0.1$, $b = 0.2$ and $c = 0.3$ (d) $a = 0.5$, $b = -0.1$ and $c = 0$

2 In Rugby Union the formula to calculate the total score is $S = 5t + 2c + 3g$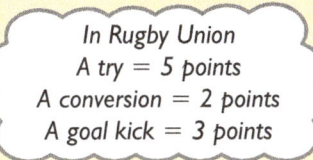

(a) What do S, t, c and g stand for?

(b) What is the total score when

    (i) $t = 2$, $c = 1$ and $g = 8$

    (ii) $t = 10$, $c = 7$ and $g = 12$

(c) In one match, England have
$t = 2$, $c = 1$, $g = 4$
and Tonga have
$t = 3$, $c = 3$, $g = 1$. Who wins?

> In Rugby Union
> A try = 5 points
> A conversion = 2 points
> A goal kick = 3 points

> Remember x^2
> means $x \times x$

3 Find the value of $5x^2 + 3y$ when

(a) $x = 2$ and $y = 3$ (b) $x = 3$ and $y = 7$

(c) $x = 10$ and $y = 30$ (d) $x = 5$ and $y = 10$

4 Find the value of $7h(h + g)$ when

(a) $h = 3$ and $g = 2$ (b) $h = 5$ and $g = -2$

(c) $h = 0$ and $g = 37$ (d) $h = 10$ and $g = -3$

5 Kerry drives for two and a quarter hours on the
motorway at 100 km per hour.

Use the formula   Distance = Speed × Time

to work out how far she has travelled.

6 (a) Martin wants to grow some vegetables in his greenhouse.
He keeps the temperature at 84°F.

Use the formula $C = \dfrac{5(F - 32)}{9}$ to convert this to Celsius.

(b) Convert the following temperatures to Celsius.

    (i) 20°F   (ii) −10°F   (iii) 0°F

(c) What is −40°F in Celsius? What is special about your answer?

7 Look at this rectangle.
Write down a formula for

(a) the perimeter of the rectangle

(b) the area of the rectangle.

The rectangle is skewed to make a parallelogram.

(c) Write down a formula for the area
of a parallelogram.

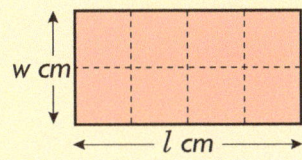

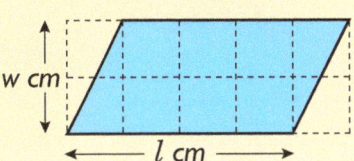

Using formulae to solve problems

Katy works for a food company.

She is designing a new cereal box.

Katy we need a new cereal box.

For each box she needs to know

- How much it will hold. This is the volume. $V = lwh$
- How much cardboard you need to make it. This is the surface area A.

Look at Katy's box.

 What are the areas, in cm², of
(a) the green face
(b) the blue face
(c) the red face?
How many faces has the box got all together?

$h = 40$ cm

$l = 25$ cm $w = 8$ cm

The surface area is given by the formula $A = 2lh + 2wh + 2lw$

 Explain this formula.

Task

1 Show that Katy's box has volume 8000 cm³.

2 Work out the surface area of Katy's box.

3 Design some different boxes with volume 8000 cm³.

4 Find the whole number values of l, w and h which give the smallest value of A when $V = 8000$.

 An open box has a square base of length *l* and height *h*.
What are the formulae for
(a) the volume   (b) the surface area of this box?

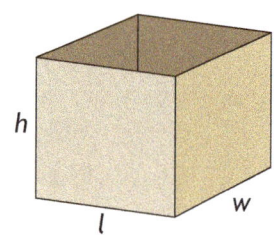

h

l w

Exercise

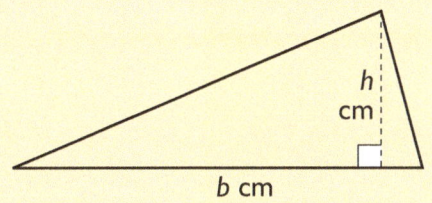

1 **(a)** Write down the formula for finding the area, A cm^2, of this triangle.

  (b) Find h when $b = 6$ and $A = 18$.

b cm / *h* cm

2 The profit a company makes is given by

> Profit = Sales − Costs

  (a) Write this as a formula. Use p for profit, s for sales and c for costs, in €.
  (b) One week the sales are €4617 and the costs €2904.
    What is the profit?
  (c) Another week $p = 2325$, $s = 4234$.
    What is c?
  (d) What does it mean if the profit is negative?

3 The change from €20 when buying some fish and chips is given by the formula

> $C = 20 - p - 2f$

  (a) What do C, p and f stand for?
  (b) Jamie buys 3 portions of chips and 4 fish.
    How much change from €20 is he given?
  (c) Phoebe buys 5 portions of chips and some fish.
    She is given €3 change.
    How many fish portions does Phoebe buy?

Fish........ *€2 each*
Portion of chips.... *€1*

4 Jamie is organising a party.
He buys some invitation cards from the local printers.
The printers work out the cost of the cards using the formula

> Cost (in cent) = 25 × number of cards + 50

This can be written as

> $C = 25n + 50$

Jamie's bill is €8.
  (a) What value should you use for C?
  (b) How many people does Jamie invite?
At the same printers Stephanie's invitations cost €16.
  (c) How many people does Stephanie invite to her party?

5 The formula for calculating the density in grams per cubic centimetres (g/cm^3) of an object is:

> Density = mass (in g) ÷ volume (in cm^3)

Glass has a density of 2.5 g/cm^3.
  (a) What is the volume of a glass ornament of mass 50 g?
  (b) Show clearly how you have checked that you have the right answer.

Finishing off

Now that you have finished this chapter you should be able to:

- understand what a formula is
- use formulae expressed in words and symbols
- substitute numbers into a formula
- derive formulae
- use formulae to solve problems.

Review exercise

1 Find the value of $2x + 3y - 5z$ when

   (a) $x = 2, y = 3$ and $z = 1$ **(b)** $x = 10, y = -20$ and $z = 50$

   (c) $x = 1, y = 1$ and $z = -1$ **(d)** $x = 0.5, y = 2$ and $z = 0$

2 Find the value of $4a^2 + 3b$ when

   (a) $a = 5$ and $b = 10$ **(b)** $a = 1$ and $b = 1$

   (c) $a = 10$ and $b = 100$ **(d)** $a = 3$ and $b = 2$

3 Find the value of $5p(2r - s)$ when

   (a) $p = 2, r = 3$ and $s = 5$ **(b)** $p = 1, r = 1$ and $s = 1$

   (c) $p = 0, r = 13$ and $s = 6$ **(d)** $p = 17, r = 2$ and $s = 4$

4 The speed limit on the motorway is 70 mph.
The police are passed by a car travelling at 110 km per hour.
The police use the formula

$$\text{Speed in kilometres per hour} = \frac{8 \times \text{speed in miles per hour}}{5}$$

to convert between miles per hour and kilometres per hour.

Was the driver over the speed limit?

5 The total resistance in ohms of the circuit, R is found by using the formulae

$$R_T = R_1 + R_2$$

The total resistance is 22.6 ohms.
What is the value of R_1?

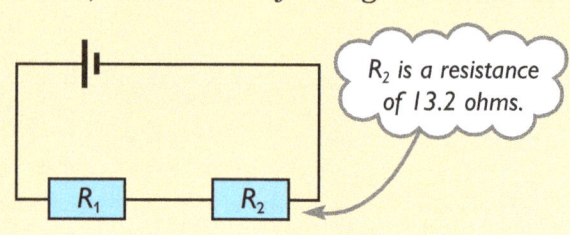

R_2 is a resistance of 13.2 ohms.

6 **(a)** Write down the formula for the area of this parallelogram.

The parallelogram is divided exactly into half.

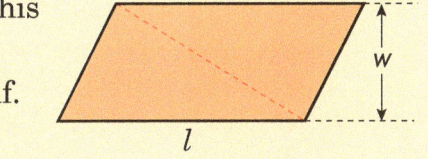

(b) Use your formula for the area of a parallelogram to find a formula for the area of a triangle.

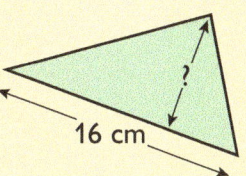

The area of a triangle is often given as
$$\text{Area} = \tfrac{1}{2}\,\text{Base} \times \text{Height}$$
$$\text{or } A = \tfrac{1}{2}\,b\,h$$
b means the same as l
h means the same as w

(c) The area of a triangle is $24\ \text{cm}^2$ and it has a base of 16 cm.
What is the height of the triangle?

16 cm

7 This stick is made out of 6 cubes. It is on a table.
(a) How many faces is it possible to see?
(b) How many faces are hidden?
Show clearly how you have found your answers.

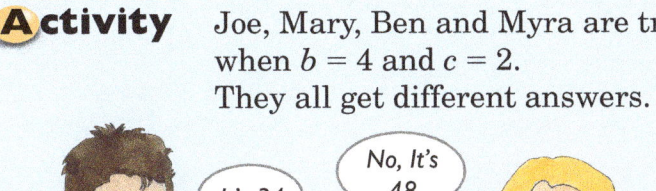

Find a formula to calculate
(c) the number of faces showing in a stick n cubes long.
(d) the number of faces hidden in a stick n cubes long.

8 Jenny orders some shirts by post for the school hockey team.
They cost €9 per shirt plus €5 overall postage.
(a) Show how to calculate the cost of 8 shirts.
(b) Write down a formula for the cost of n shirts.
(c) Use your formula to find the cost of 10 shirts.
(d) How many shirts were ordered when the total price was
(i) €86   **(ii)** €68   **(iii)** €113?

Activity Joe, Mary, Ben and Myra are trying to find the value of $5b^2 - 3c$
when $b = 4$ and $c = 2$.
They all get different answers.

It's 34

No, It's 48

No, It's 74

You are all wrong. It's 2884

Joe

Mary

Ben

Myra

Who has the right answer?
Explain where the others have gone wrong.

Perimeter, area and volume

Eric's silly morning

After his breakfast, at eighteen-oh-five,
Eric jumps into his car,
Nineteen square metres the distance to drive,
It's not exactly too far.

A three litre parking space, easy to fit,
Four kilograms from the front door,
Ten metres in volume, is the smart lift,
But what is the size of its floor?

The top of his desk is rectangular,
Three metres, the length of one side,
And with Eric's great mountain of paper,
It has to be two litres wide.

 What is silly about Eric's morning?

Task

1   Rewrite the rhyme as a sensible *story*.
2   What is **(a)** the area **(b)** the perimeter of Eric's desktop.
3   Eric's office has a rectangular floor of area 48 m². Copy and complete this table of possible **dimensions** for Eric's office.

Area	Length	Width	Perimeter
48 m²	8 m	6 m	28 m
48 m²	12 m		
48 m²		6.4 m	
48 m²			38 m

4   The lift floor is square with an area of 4 m². The volume of the lift is 10 m³.
   (a) What are the length and width of the floor of the lift?
   (b) What is the height of the lift?

5   Eric has a new computer fitted in the corner of his office. Here is a plan view.
   (a) Calculate the perimeter.
   (b) Calculate the area.

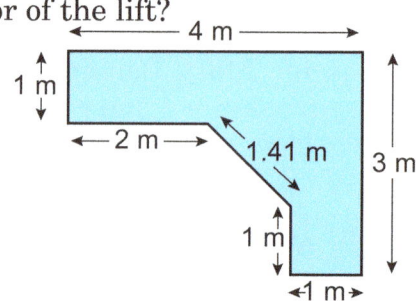

Perimeter is the distance around a flat shape
Perimeter is measured in length units eg cm
Perimeter of a rectangle = 2 × (*l* + *w*)

Area of a rectangle = length × width
Area is measured in square units eg cm²

Area of a triangle = $\frac{1}{2}$ × base × height

Volume of a cuboid = length × width × height
Volume is measured in cubic units eg cm³

Exercise

1   Calculate the areas of these rectangles.

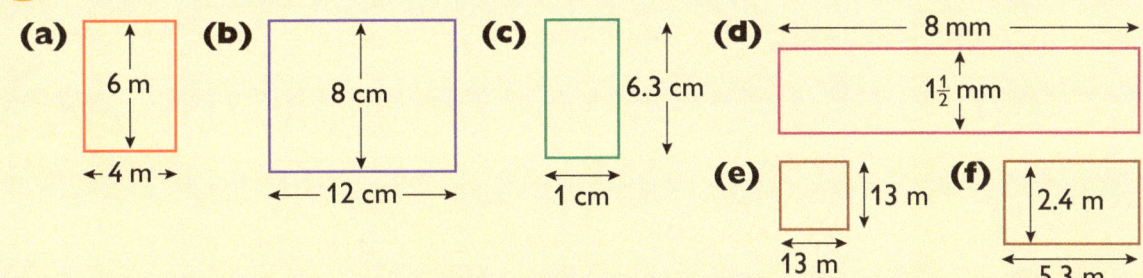

(a) 6 m, ←4 m→   **(b)** 8 cm, ←12 cm→   **(c)** 6.3 cm, 1 cm   **(d)** 8 mm, $1\frac{1}{2}$ mm   **(e)** 13 m, 13 m   **(f)** 2.4 m, 5.3 m

2   Calculate the areas of these triangles.

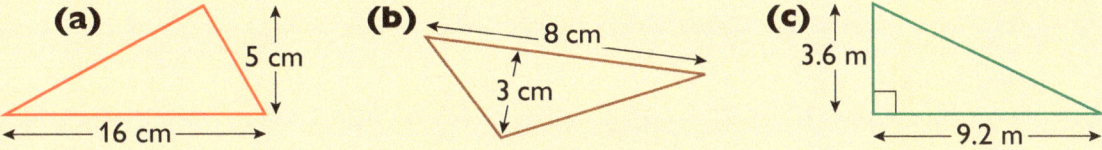

(a) 5 cm, 16 cm   **(b)** 8 cm, 3 cm   **(c)** 3.6 m, 9.2 m

3   Calculate the perimeters of these shapes.

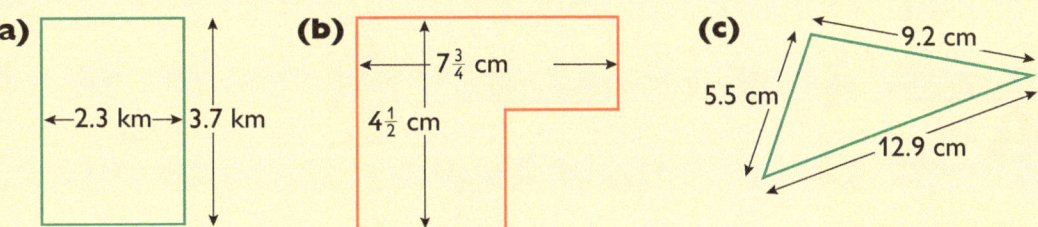

(a) 2.3 km, 3.7 km   **(b)** $7\frac{3}{4}$ cm, $4\frac{1}{2}$ cm   **(c)** 9.2 cm, 5.5 cm, 12.9 cm

4   Calculate the volumes of these cuboids.

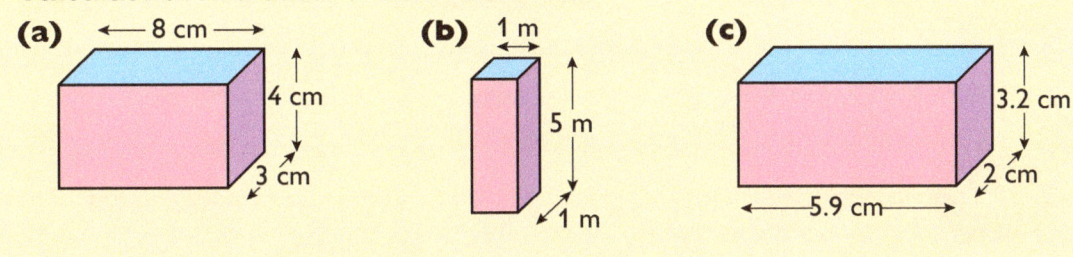

(a) 8 cm, 4 cm, 3 cm   **(b)** 1 m, 5 m, 1 m   **(c)** 3.2 cm, 2 cm, 5.9 cm

5   Calculate **(i)** the perimeter and **(ii)** the area of

(a)   a rectangle measuring 4.2 cm by 5 cm

(b)   a square with sides of length 6.3 cm

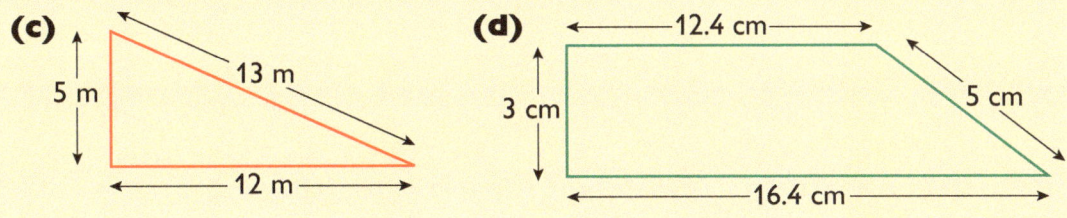

(c) 5 m, 13 m, 12 m   **(d)** 12.4 cm, 3 cm, 5 cm, 16.4 cm

Activity   **1**   Write your own silly story like Eric's, with incorrect units.
2   Swap with a friend and correct each other's story.
Make sure you agree on the correct versions!

More areas

The full formula for the area of a triangle is

> Area of a triangle $= \frac{1}{2} \times$ base $\times$ perpendicular height

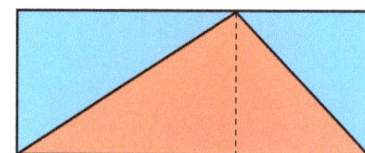

  How does the diagram explain the formula?
Why is it called *perpendicular* height?

Task

1 Work out the area of these parallelograms.

(a)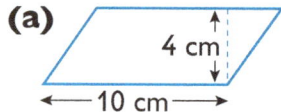

(b)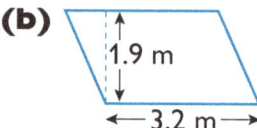

2 Write down the formula for the area of this parallelogram.
It should involve b and h.
Draw diagrams to explain why your formula works.
Explain *exactly* what b and h stand for.

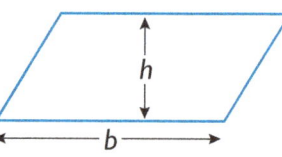

 This parallelogram uses different letters from the one on page 135.

Here are two congruent trapezia.
They can be joined to form a parallelogram.

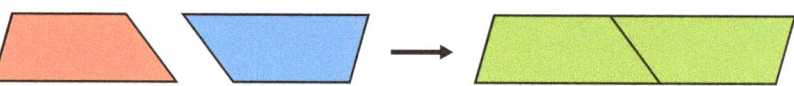

  Does this work for any pair of congruent trapezia?

Task

1 Find the area of this parallelogram.

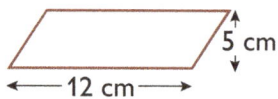

2 What is the area of each trapezium in this diagram?

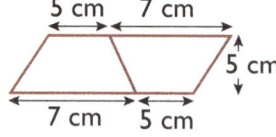

3 Work out the formula for the area of this trapezium. It should involve a, b and h.

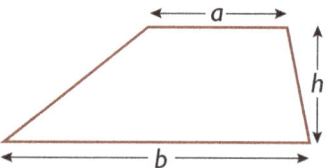

Draw diagrams to explain why your formula works.
Explain *exactly* what a, b and h stand for.

Any polygon can be split into rectangles, triangles and trapezia to find its total area.

Exercise

1 Use the appropriate formula to calculate the areas of these shapes.

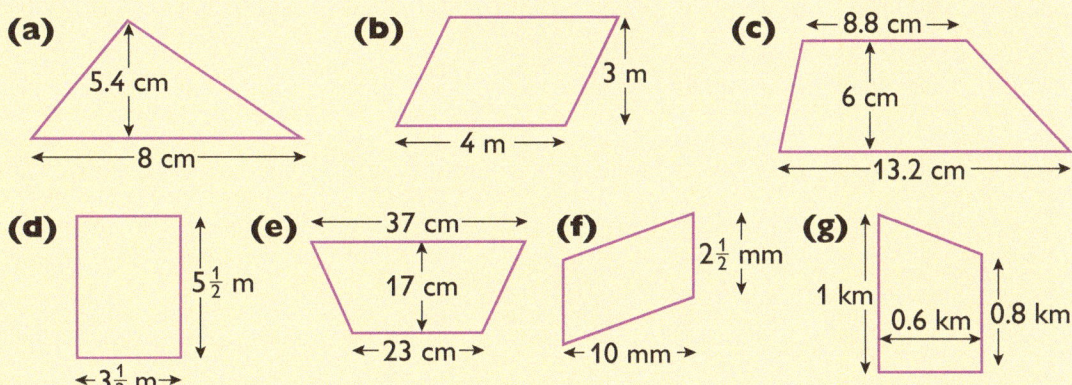

(a) 5.4 cm, 8 cm

(b) 3 m, 4 m

(c) ←8.8 cm→, 6 cm, ←13.2 cm→

(d) $5\frac{1}{2}$ m, $3\frac{1}{2}$ m→

(e) ←37 cm→, 17 cm, ←23 cm→

(f) $2\frac{1}{2}$ mm, ←10 mm→

(g) 1 km, 0.6 km, 0.8 km

2 Here is a plan of Megan's garden.

   (a) Fencing costs €1.20 per metre.
      Calculate the cost to fence
      around the garden.

   (b) Megan decides to sow the
      garden with lawn seed.
      One packet of seed
      covers 12.5 m².
      How many packets will she need?

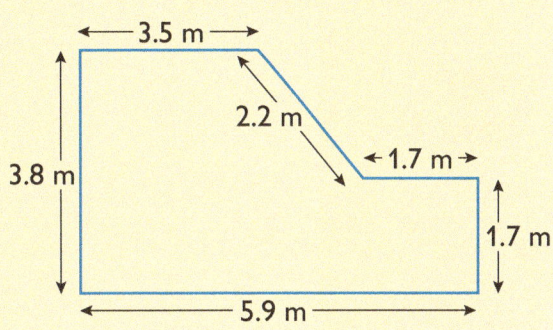

3 The diagram shows two ornamental
ponds surrounded by concrete.

   (a) Calculate the area of concrete.

   Concrete costs €2.70 per
   square metre.

   (b) Calculate the cost of the concrete
      surrounding the ponds.

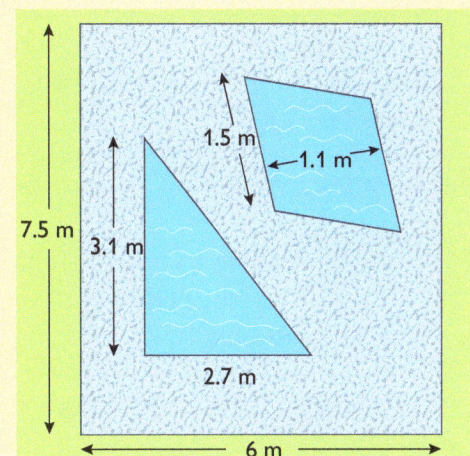

4 This diagram shows two squares.
The measurements are in metres.
Think about the areas of the squares.
Work out the length l.
Give your answer correct to the nearest centimetre.

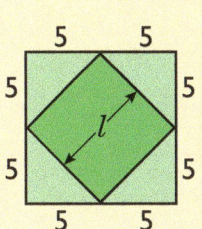

Volume and surface area

Here is a **net** of a cuboid.
Each rectangle makes a face of the cuboid.
The shaded rectangle is the bottom face.

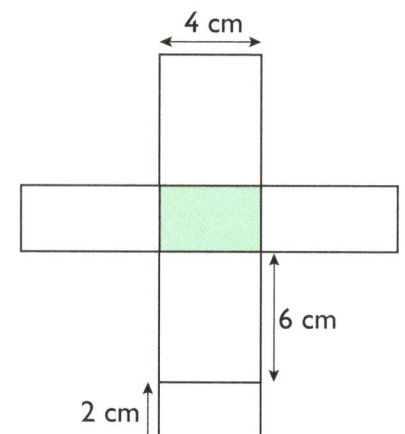

 What are the *dimensions* of the cuboid?
What is the volume of the cuboid?
How many faces does the cuboid have?
How many *different* faces does the cuboid have?
What can you say about opposite faces of a cuboid?

 What is the area of the top face?
What is the area of the net?
How do you work out the area of the net?
What is the *surface area* of the cuboid?

Task

Look at this open cardboard box. It has no top.

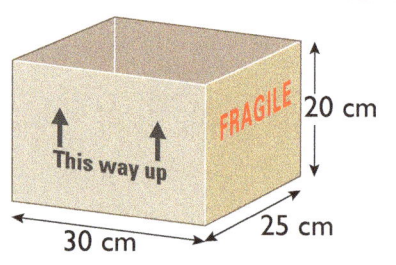

1 Draw the net of the box, using a scale of 1 : 5.

 Why do you need to draw the net to scale?
What is the problem with drawing the net full size?

2 Copy and complete this table to find the area of cardboard.

Area of bottom	= 30 cm × 25 cm	Area of bottom	=	30 cm × 25 cm	= 750 cm²
Area of front	= 30 cm × 20 cm	Area of front and back	= 2 ×	×	=
Area of right side	=	Area of both sides	=		=
		Total surface area	=		= cm²

3 Make a table to calculate the
area of wood used to make
this open box.

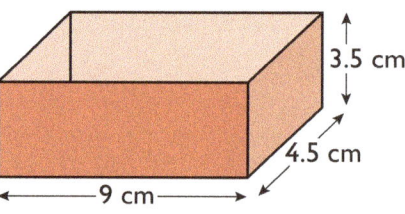

4 Write down the formula for the surface area of the
outside of this open box.
It should involve l, w and h.
Explain what l, w and h stand for.

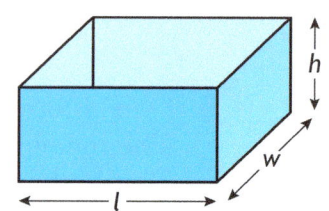

 Which is the easiest way to find the surface area of a cuboid?
Is it the formula, or by using a net?

Exercise

1 Calculate **(i)** the volume and **(ii)** the surface area of the following cuboids.

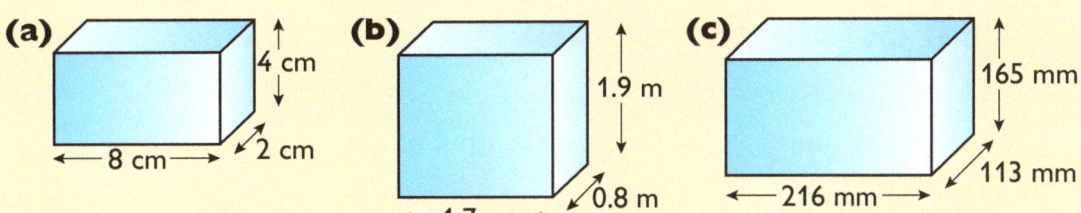

(a) 4 cm, 8 cm, 2 cm

(b) 1.9 m, 1.7 m, 0.8 m

(c) 165 mm, 216 mm, 113 mm

(d) A cuboid with length 4.34 m, width 1.32 m and height 3.4 m.

2 This L-shaped prism is made by putting together two cuboids.
Calculate

(a) the volume of each cuboid

(b) the volume of the prism

(c) the surface area of each cuboid

(d) the surface area of the prism.

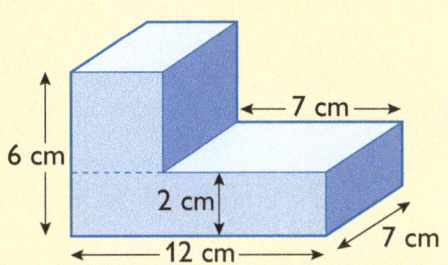

6 cm, 7 cm, 2 cm, 12 cm, 7 cm

3 **(a)** Calculate the volume of this C-shaped prism by splitting it into 3 cuboids. There are different ways to do this.

(b) Calculate the surface area of the prism.

The prism is made of wood with a density of 0.74 grams per cubic centimetre.

(c) Calculate the mass of the prism.

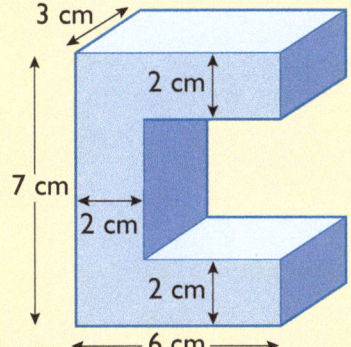

3 cm, 2 cm, 7 cm, 2 cm, 2 cm, 6 cm

4 You can calculate the volume of a prism using the formula

> Volume of prism = Area of cross section × Length

Use this formula to calculate the volume of the prism in Questions 2 and 3.
Check your answer.

Investigation

1 What is the formula for the surface area of a cube of width x cm?

2 Find the surface area of a cuboid made by putting together two cubes as shown.

3 What is the surface area of a cuboid made from four cubes in a line?

4 Find the surface area of a cuboid made from n cubes in a line.

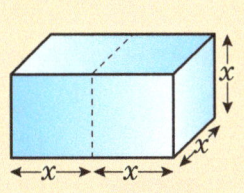

x, x, x, x

Finishing off

Now that you have finished this chapter you should know:

- the correct units for length, area and volume
- that the perimeter is the distance around a flat shape
- these formulae:

  Area of a rectangle = length × width
  Perimeter of a rectangle = $2 \times (l + w)$
  Area of a triangle = $\frac{1}{2}$ × base × perpendicular height
  Area of a parallelogram = base × perpendicular height
  Volume of a cuboid = length × width × height
  Surface area of a cuboid = $2lw + 2lh + 2wh$

  Area of a trapezium
  = $\frac{1}{2} \times (a + b) \times h$

Review exercise

1 Calculate the area of each of these shapes.

(a) 11 cm, 13 cm

(b) 23 cm, 1 cm

(c) 3.6 m, 5.3 m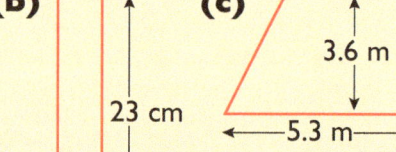

(d) 3 km, 3 km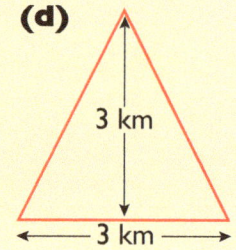

(e) 5.4 cm, 2.1 cm, 2.6 cm

(f) 5 cm, 5 cm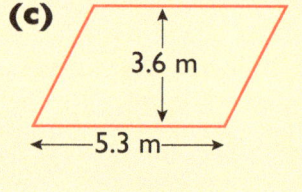

(g) 1.7 m, 0.7 m, 2.7 m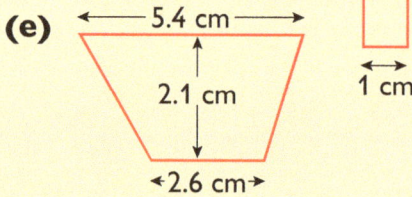

(h) 103 mm, 35 mm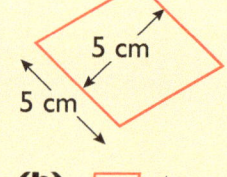

(i) $5\frac{1}{2}$ cm, $2\frac{3}{4}$ cm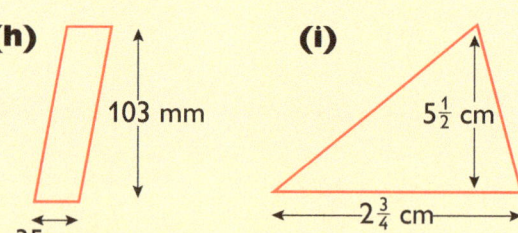

2 For each of the measurements below choose appropriate metric units.

Example The area of a floor would be measured in m².

(a) length of a car
(b) area of a desk top
(c) volume of a filing cabinet
(d) surface area of a matchbox
(e) volume of a matchbox
(f) distance across the English Channel
(g) thickness of a match
(h) height of a tree
(i) area of a wall
(j) area of a blade of grass
(k) volume of a match
(l) volume of a shoe box.

3 Calculate **(i)** the volume and **(ii)** the surface area of cuboids with the following dimensions:

(a) length 5 cm, width 3 cm, height 4 cm

(b) length 4 cm, width 5 cm, height 3 cm

(c) $7\frac{1}{2}$ m by $3\frac{1}{2}$ m by $10\frac{1}{2}$ m

(d) 5 mm by 5 mm by 5 mm

4 Match each of the seven formulae at the top of the opposite page with its appropriate units.

Example Area of rectangle = length × width → square units

5 Avonshire Water dig a cuboid-shaped hole in the road.
The hole is 3.5 m long, 2.3 m wide, and 2.7 m deep.

(a) What is the perimeter of the top of the hole?

The workmen put up a rectangular fence around the hole.
The fence is 1 m from the edge of the hole.

(b) Calculate the perimeter of the fence.

A cuboid-shaped tank is placed in the hole.
It measures 2.1 m by 1.8 m by 1.5 m.

(c) Calculate the volume of rubble needed to fill in the rest of the hole.

6 An open topped glass fish tank is cuboid-shaped.
It is 50 cm long, 30 cm wide and 35 cm high.

(a) How much glass is used to make the tank?

The tank is filled with water up to a depth
of 34 cm.

(b) How many litres of water are used?

A brick is placed in the tank.
The brick measures 21 cm by 10 cm by 6 cm.

(c) Does any water spill out of the tank?

7 Floppy disks measure 89 mm by 93 mm by 3 mm.
A box is made to contain a stack of 20 disks.

(a) Calculate the dimensions of the box in *centimetres*.

(b) Calculate the surface area of the box.

(c) Calculate the volume of the box.

Investigation

Susie is making a patio in her garden.
She has 144 paving slabs.
Each slab is square, with a width of two feet.
She uses all 144 slabs to make a rectangular patio.

1 Find how many different sized rectangular patios she can make.

2 Which patios have the greatest and least perimeters?

3 What can you say about the areas of the patios?

18 Percentages

 There are 600 boys and 600 girls at Avonford High School. How many do sports? How many do dancing?

Task

Here is a chart for converting between percentages, decimals and fractions. Make your own copy, 20 cm long and 8 cm wide. (You can use graph paper.) Fill in the missing numbers.

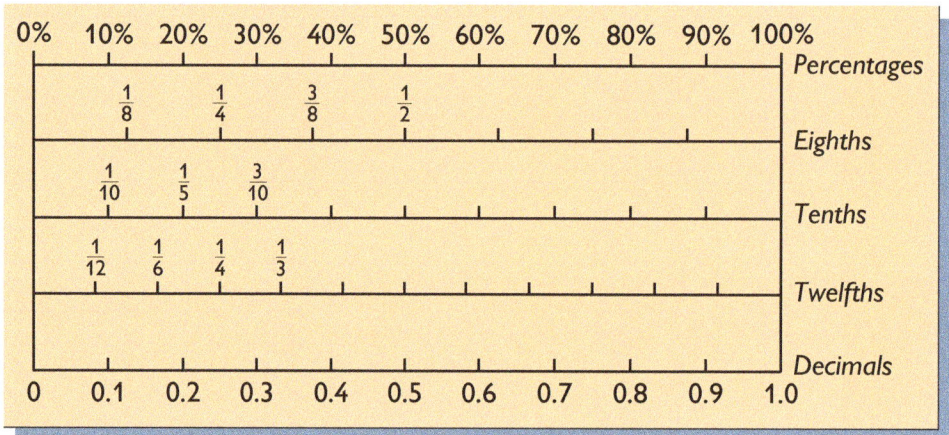

Set your friends some questions to do with your chart.

For example, 'Roughly, what is $\frac{1}{12}$ as a percentage?'

Remember that percentage means out of 100

42% can be written as a fraction $42\% = \frac{42}{100} = \frac{21}{50}$ *(Cancelling by 2.)* or as a decimal $42\% = 0.42$

Similarly $28\% = \frac{28}{100} = \frac{7}{25}$ *(Cancelling by 4.)* or as a decimal $28\% = 0.28$

 In 0.42 the 4 means four tenths. What does the 2 mean?

Remember to change fractions or decimals to percentages multiply by 100% (a whole one)

$$\frac{2}{5} = \frac{2}{5} \times 100\% = 40\% \qquad 0.37 = 0.37 \times 100\% = 37\%$$

Exercise

1 Convert these percentages to fractions.

(a) 50%   (b) 25%   (c) 60%   (d) 10%   (e) 24%

(f) 85%   (g) 4%   (h) 5%   (i) 16%   (j) 3%

2 Convert these percentages to decimals.

(a) 75%   (b) 30%   (c) 63%   (d) 7%   (e) 12.5%

(f) 1%   (g) 2.5%   (h) 6.5%   (i) 0.1%   (j) $\frac{1}{2}$%

3 Convert these fractions and decimals to percentages.

(a) $\frac{3}{5}$ (b) $\frac{4}{25}$ (c) $\frac{11}{20}$ (d) $\frac{7}{10}$

(e) $\frac{5}{9}$ (f) $\frac{3}{7}$ (g) $\frac{5}{8}$ (h) $\frac{7}{11}$

(i) 0.11   (j) 0.06   (k) 0.4   (l) 0.375

4 Write these fractions as percentages.

(a) $\frac{1}{4}$ (c) $\frac{3}{4}$ (c) $\frac{2}{5}$ (d) $\frac{1}{3}$ (e) $\frac{2}{3}$

Activity **1** Copy and complete this table.

Fraction	Decimal	Percentage
$\frac{1}{2}$		
	0.25	
		75%
$\frac{1}{3}$		
	0.6	
	0.1	
		20%

 Do you think it is useful to learn this table? Why?

2 Look at this example. It shows how to convert $12\frac{1}{2}$% to a fraction.

$$12\frac{1}{2}\% = \frac{12\frac{1}{2}}{100} \times \frac{2}{2}$$

Multiply top and bottom by 2 to get rid of the half on the top line

$$= \frac{25}{200} = \frac{1}{8}$$

 Change these percentages to fractions without using a calculator.

(a) $37\frac{1}{2}$%   (b) $87\frac{1}{2}$%   (c) $62\frac{1}{2}$%

More percentages

THE AVONFORD STAR

Major Fish Catch is Wet Fish

One year 90% of the total Avonmouth harbour catch of 800 tonnes was wet fish. The rest was shellfish.

Look at the article.

$$90\% \text{ of } 800 \text{ tonnes} = \frac{90}{100} \times 800 \text{ tonnes}$$
$$= 720 \text{ tonnes}$$

90% is wet fish. So 10% is shellfish.

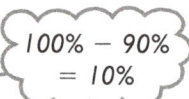

100% − 90% = 10%

? **What weight of shellfish is this?**

Task

The catch of wet fish is shown in the bar chart opposite.

(a) Copy and complete the table.

Fish	%	Weight in tonnes
Cod	15%	108
Haddock		
Plaice		
Coalfish (Saithe)		
Whiting		
Herring		
Mackerel		

(b) Check that the total percentage is 100%.
Check that the total weight is 720 tonnes.

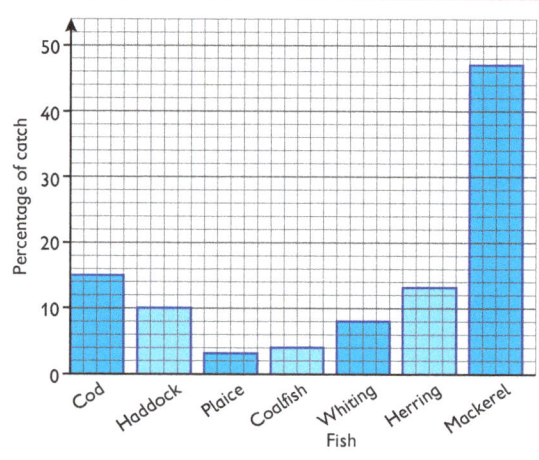

(c) Draw a pie chart to show the information in the bar chart.

Notice that **720 tonnes is 360° (the total)**

so **1 tonne is $\frac{360°}{720} = \frac{1}{2}°$**

Cod is 108 tonnes, so the angle is
$108 \times \frac{1}{2}° = 54°$

0.36 tonnes out of the 72 tonnes of haddock is rejected as unsuitable.

? **What percentage is rejected?**

People often say this as 'half of one percent'.

Fraction rejected $= \frac{0.36}{72} = \frac{1}{200}$ Percentage rejected $= \frac{1}{200} \times 100\% = \frac{1}{2}\%$ (or 0.5%).

Exercise

1 A sweater is made from 70% acrylic. The rest is wool.
What percentage is wool?

2 A skirt is 60% lambswool, 20% angora and the rest is polyamide.
What percentage is polyamide?

3 Brass is made up of $66\frac{2}{3}\%$ copper. The rest is zinc.
What percentage is zinc?

4 **(a)** Stella scores 64 marks out of a possible 80 in a French test.
What is this as a percentage?

(b) Simon scores 21 out of 60 in an English test.
What is his percentage?

5 Jane does a survey of the pet cats in her village.

Black and/or white	Tabby	Ginger	Tortoiseshell	Siamese	Others
36	42	24	3	6	9

(a) Write these figures as percentages.

(b) Draw a pie chart to illustrate these data.

Investigation These pie charts show the proportion of food types in certain foods.

(a) Cheese

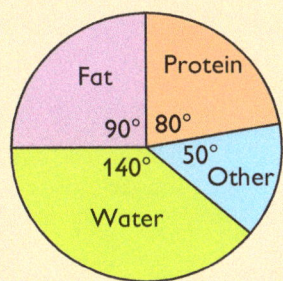

Copy and complete the answer to the question
'What percentage of cheese is fat?'

Fraction of fat $= \frac{90}{360}$, % of fat $= \frac{90}{360} \times 100 = \ldots$

(b) Streaky bacon

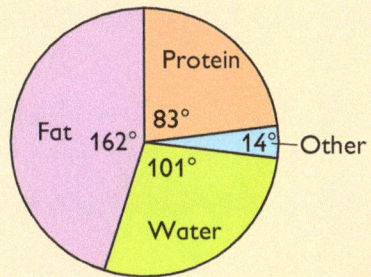

What percentage is fat?

(c) Eggs

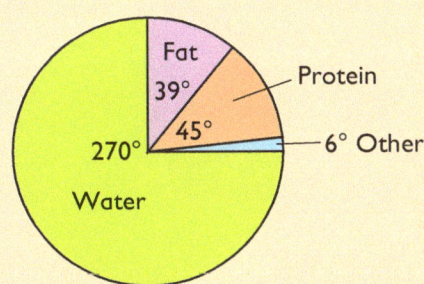

What percentage is
(i) protein? **(ii)** water?

(d) Wholemeal bread

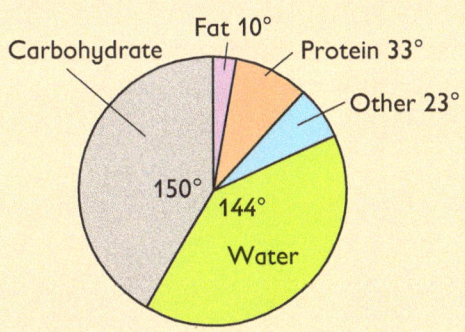

What percentage is
(i) carbohydrate? **(ii)** fat?

Everyday percentages

AVONFORD TIMES

Local nurses

Nurses at Avonford Hospital say they need a 4% pay increase to meet the rise in the cost of living.

AVONFORD TIMES

Avonford nurses accept 3% increase

This is less than the 4% they had wanted.

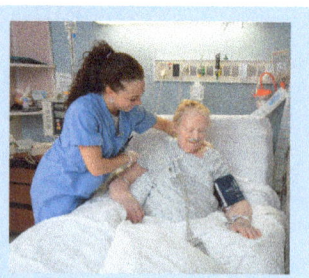

Wanda works in the Accident and Emergency Department of Avonford Hospital. She was earning €21 000 before the rise.

She writes

> 3% of 21 000 $= \frac{3}{100} \times 21\,000$
> $= €630$
> New salary $= €(21\,000 + 630) = €21\,630$

 Wanda is not happy with the 3%. She says '4% is much more'. How much more is 4% for Wanda?

Wanda's friend Spike says he can work it out quicker.

He writes

> $100\% + 3\% = 103\%$
> 103% of €21 000
> $\frac{103}{100} \times €21\,000 = €21\,630$

 Which method is easier, Wanda's or Spike's?

Make a poster showing everyday use of percentages.
Possible sources are newspapers, magazines, leaflets and brochures.

Sometimes you need to work out a percentage decrease.
Jon bought a new bicycle 3 years ago. It costs €400.
Now it has lost 30% of its value. What is its value now?

> 30% of €400 $= \frac{30}{100} \times €400$
> $= €120$
> Its value is €400 − €120 = €280

 Another way of working this starts 100% − 30% = 70%. How do you continue?

Exercise

1 The Government awards a 3% pay increase to all public sector workers.
Work out the new salary for
  (a) Beatrice who earned €15 000   **(b)** Brett who earned €24 000
  (c) Bert who earned €17 000.

2 Cars depreciate in value as they get older.
What is the value after one year's depreciation of these cars
  (a) Fiat Brava originally €8750, depreciation 20%
  (b) Peugeot 406 originally €13 750, depreciation 20%
  (c) Peugeot 806 originally €17 500, depreciation 9%
  (d) Chrysler Neon originally €13 495, depreciation 25%
  (e) Vauxhall Astra originally €16 140, depreciation $16\frac{2}{3}$%?

3

SALE 10% off

Find the sale price of these items.
  (a) Shoes   €30   **(b)** Jeans   €55   **(c)** Jumper   €42

4 Value-added tax is charged at $17\frac{1}{2}$%.
In each of these cases find
      (i) the VAT   **(ii)** the total.
  (a) A restaurant bill of €50 + VAT
  (b) A garage bill of €70 + VAT
  (c) A telephone bill of €35 + VAT
  (d) A video-recorder priced €450 + VAT
  (e) A TV priced €250 + VAT.

5 32% of a tropical fruit drink is made from juices of orange, lemon,
pineapple, and apricot.
How much of these juices is there in
  (a) a 1 litre bottle   **(b)** a 350 ml glass   **(c)** a 2.5 litre jug?

6 Alvin buys a guitar. Its price is €400. He pays for it in instalments.
First he pays a deposit of 35%.
  (a) How much deposit does he pay?

Then he makes 12 monthly payments.
Each of these is 8% of the price.
  (b) What is
        (i) each monthly payment
        (ii) the total of all 12 monthly payments?
  (c) How much does he pay altogether?
  (d) How much more than €400 is this?

Finishing off

Review exercise

1 Gail scores the following marks in tests.
Convert them to percentages.
(a) French $\frac{13}{20}$ **(b)** Maths $\frac{17}{25}$ **(c)** RE $\frac{7}{10}$ **(d)** Science $\frac{37}{40}$ **(e)** IT $\frac{39}{65}$

2 Convert these percentages to fractions in their lowest terms.
(a) Sale 20% off **(b)** Members: 30% discount
(c) Music exam: distinction 80% **(d)** Deposit of $12\frac{1}{2}\%$ required

3 Convert these decimals to percentages.
(a) 0.4 **(b)** 0.22 **(c)** 0.06 **(d)** 0.025

4 44 people out of 80 buy a programme at a rugby match.
What percentage of the people buy a programme?

5 A shop receives 700 light bulbs. 1% of them are faulty.
How many of the light bulbs are faulty?

6 A shop buys 600 batteries.
$\frac{1}{2}\%$ of them are faulty.
How many batteries are faulty?

7 A tub of Fruit Fool consists of 8% lemon.
The tub contains 115 g of Fruit Fool.
How many grams of lemon are there in a tub?

8 David buys a keyboard priced at €450.
He pays the shopkeeper a deposit
of 35% of the price of the keyboard.
How much deposit does he pay?

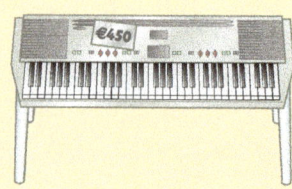

9 Ellen buys a guitar priced at €150. First she pays a deposit of 30%.

(a) How much deposit does she pay?

(b) She then pays 12 monthly instalments of €11 each.
How much do the instalments come to altogether?

(c) What is the total of the deposit and the 12 monthly instalments?

(d) How much extra does she pay for the guitar?

10 18 out of 60 calculators have flat batteries after two years.
What percentage is this?

11 6 out of 30 pupils attained full marks in their mathematics test.
What percentage of pupils is this?

12 3 textbooks out of 150 are discarded at the end of the school year because they are falling apart.
What percentage is this?

13 The price of the following items goes up by 20%.
What is the new price of each item?

(a) CD €15

(b) Jeans €32

(c) Video recorder €280

(d) Theatre ticket €12.50

14 The value of these items has increased.
Find their new values.

(a) Old postage stamp €5, increased by 20%

(b) Antique pottery €150, increased by 65%

(c) Diamond ring €1500, increased by 124%

(d) Vintage car €1800, increased by 26%

15 The value of these items has gone down.
What is the new value of each item?

(a) Bicycle €300, down by 10%

(b) Guitar €250, down by 15%

(c) Model car €30, down by 75%

16 (a) 32 out of the 400 people on a train travelled first class.

(i) What percentage of people travelled first class?
One of the people on the train is chosen at random for a prize.

(ii) What is the probability that the prize-winner travels first class?

(b) 700 people are given an injection.
49 of them develop temperatures.

(i) What percentage of the people do not develop a temperature?

(ii) Estimate the probability that a person will not develop a temperature after this injection.

19 Symmetry

Do you remember?

? How many lines of reflection symmetry does this club symbol have?
Why does it have *no* rotational symmetry?

? What is the order of rotational symmetry of this flash symbol?
Why does it *not* have reflection symmetry?

? What is the order of rotational symmetry of this diamond symbol?
How many lines of symmetry does it have?
Where is the centre of rotational symmetry?

Task

1 Look at these shapes drawn on squared paper.

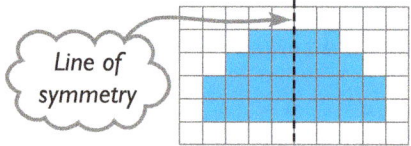

Line of symmetry

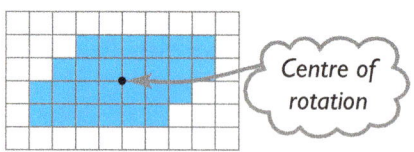

Centre of rotation

Shape 1 has one line of symmetry.
It has no rotational symmetry.

Shape 2 has rotational symmetry order 2.
It has no lines of symmetry.

2 Copy Shapes 1 and 2 onto squared paper.
Use a mirror and tracing paper to check their symmetries.

3 The symmetries of Shapes 1 and 2 are shown in the table opposite.
Draw your own designs for Shapes 3, 4, 5 and 6.
Label each shape and show the lines of symmetry and centre of rotational symmetry.
Use a mirror and tracing paper to check your shapes.

	Number of lines of symmetry	Order of rotational symmetry
Shape 1	1	–
Shape 2	0	2
Shape 3	2	2
Shape 4	0	–
Shape 5	4	4
Shape 6	0	4

A shape can have reflection symmetry *or* rotational symmetry *or* both.
If it has no rotational symmetry and no reflection symmetry it is **asymmetric**.

? Shape 3 has 2 lines of symmetry and order of rotational symmetry 2.
Can you design a shape with 2 lines of symmetry and no rotational symmetry?

? A body has rotational symmetry order 1. What can you say about it?

Exercise

1 **(a)** Copy these shapes and add all the lines of symmetry to each one.

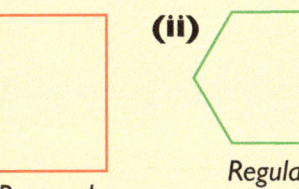

(i) Rectangle **(ii)** Regular hexagon **(iii)** Square **(iv)** Parallelogram **(v)** Kite

(b) For each shape write down the order of rotational symmetry.

2 Here are parts of shapes which have two lines of symmetry.

(a) 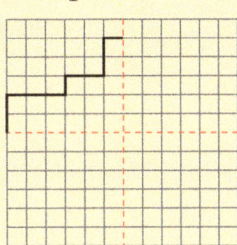 **(b)**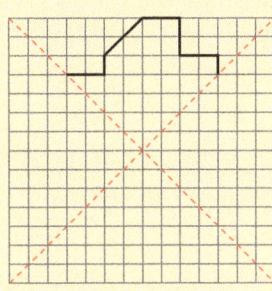

(i) Copy each one and draw in the rest of the shape.
(ii) Write down the order of rotational symmetry of each shape.

3 **(a)** Look at the diagram.
    (i) What is the shape?
    (ii) How many lines of symmetry does it have?
    (iii) What is its order of rotational symmetry?

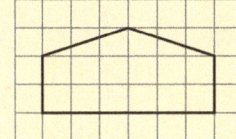

(b) Use squared paper to design a decagon with rotational symmetry order 2.

(c) Use squared paper to design a 28 sided shape with rotational symmetry order 4.

Activity

Crossword puzzles are usually square and usually symmetrical.
This one has rotational symmetry order 2.

Ask your teacher for a copy of this puzzle or copy it onto squared paper.

You are going to change the pattern so that it has rotational symmetry order 4.

Do this by shading in some more squares.
Shade as few squares as possible.

Keep a record of the number of squares you shade.
Who uses the least?

Design a 10 × 10 grid which has 4 lines of symmetry.

What is its order of rotational symmetry?

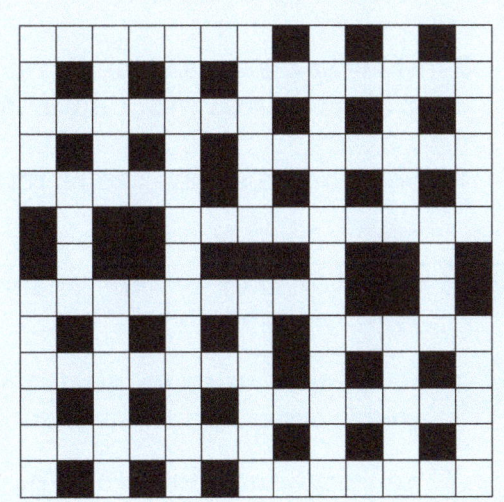

Making symmetrical shapes

In Question 2 on page 157 you made symmetrical shapes by reflecting patterns in a line of symmetry.
You can also make symmetrical shapes by rotating polygons.

Here an equilateral triangle is rotated to produce a regular hexagon.

Equilateral triangle

Rotations of 60° about a *vertex*. *Six* positions.

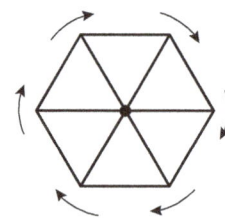

Regular hexagon
Rotational symmetry order six.

The centre of rotation is a vertex.

Task

Here a triangle is rotated to produce a rectangle.
The centre of rotation is the mid-point of a side.

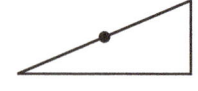

Right-angled scalene triangle

Rotate through 180° about mid-point of longest side.
Two positions.

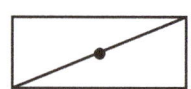

Rectangle
Rotational symmetry order 2.

You may find squared paper helpful.

I **(a)** Draw a diagram to show a right-angled scalene triangle rotated through 180° about the mid-point of one of the shorter sides.
What shape is produced?

(b) Repeat part (a) using the mid-point of the other shorter side.

2 Show how a square can be rotated to produce a rectangle.

3 Show how a square can be rotated to produce a cross with rotational symmetry order 4.

4 Show how a square can be rotated to produce a different square.

? **What sorts of triangle can be rotated to produce a rhombus? How can it be done?**

? **Find some rules for producing symmetrical shapes by rotating polygons about their vertices or mid-points.**

Your rules may involve
the order of rotational symmetry of the final shape
the number of positions
the angle of each rotation.

Exercise

1 A triangle is rotated to produce a regular octagon.

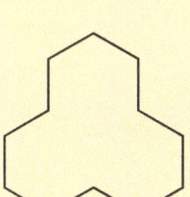

    (a) Calculate the interior angles of the triangle.

    (b) What sort of triangle is it?

    (c) Calculate the interior angles of the octagon.

2 All the obtuse angles on this dodecagon are 120°.

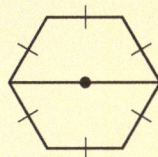

Draw a diagram to show how it can be produced by rotating a different polygon.

You may find isometric paper useful for your diagram.

3 This diagram shows a trapezium rotated to produce a regular hexagon.

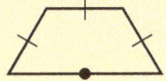

Rotation of 180° about
mid-point of base.
Two positions.

Why do three sides of the trapezium have to be equal in length?

4 Draw diagrams to show how a trapezium can be rotated to produce the following shapes.

In each case **(i)** show any ways in which the trapezium has to be special.

               (ii) describe the symmetry of the final shape.

    (a) An irregular hexagon   **(b)** A parallelogram

    (c) A rectangle   **(d)** A square

    (e) A rhombus   **(f)** Another trapezium.

Investigation

1 Draw a rectangle on squared paper.
Mark a centre of rotation which is *not* at a
mid-point, as shown.
Rotate the rectangle 4 times through 90°.
Describe the symmetry of the shape produced.

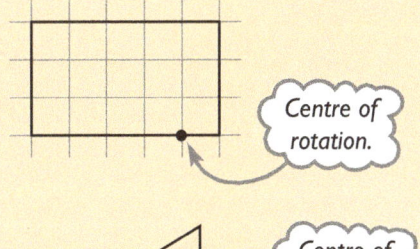

Centre of
rotation.

2 A right-angled scale triangle is rotated about
the centre shown to produce a shape with
rotational symmetry order 8.
Draw a diagram to show this.
What regular polygon have you formed?

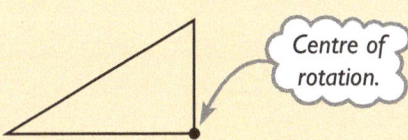

Centre of
rotation.

3 A triangle is rotated to produce a square. Show two ways of doing this.
Describe the triangle.

? **Does this investigation fit the rules on the opposite page?**

Symmetry of polygons

Here are three regular polygons
with their lines of symmetry shown.

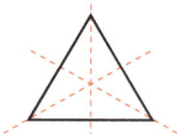 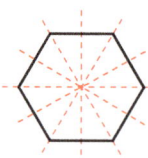 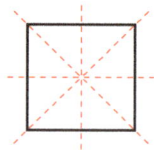

 **What can you say about the number of sides of a regular polygon,
the number of lines of symmetry and the order of rotational symmetry?
What is the connection between the lines and the centre of rotational symmetry?**

Symmetrical shapes can be produced by rotating polygons about their centres.
In this diagram an equilateral triangle is rotated to produce a six-pointed star.

Example

Equilateral triangle.
Rotational symmetry order 3.

Rotation of 180° about
centre. *Two positions.*

Six-pointed star.
Rotational symmetry order 6.

Task

1 **(a)** Construct a square with sides of 8 cm on a piece of card.
Cut out the square to use as a template.
(b) Use lines of symmetry to find the centre of rotational symmetry of the square.

2 Use your template to draw a diagram to show a
square rotated to produce a star with rotational
symmetry order 8.

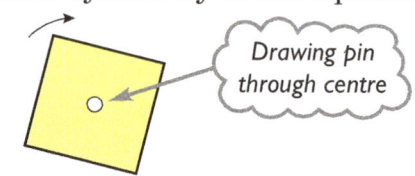

Drawing pin
through centre

3 Check the symmetry of your star by using tracing paper.

 **In how many positions is the square placed to make the eight-pointed star?
What is the angle of rotation between each square?**

4 **(a)** Use your template to draw a diagram to show how a square can be rotated to
produce a star with rotational symmetry order 12.
(b) Check the symmetry with tracing paper.

 **In how many positions is the square placed to make the twelve-pointed star?
What is the angle of rotation between each square?**

 Can you make a ten-pointed star by rotating a square?

 Find a rule for rotating polygons about their centres to produce symmetrical shapes.

Your rule may involve
the order of rotational symmetry of the polygon
the number of positions
the order of rotational symmetry of the final shape.

Exercise

1 **(a)** Show how a cross with rotational symmetry order 4 can be made by rotating a rectangle.

(b) Does this obey the rules you found on the opposite page?

2 **(a)** Construct an accurate equilateral triangle with sides of length 8 cm on some card.

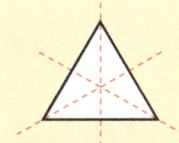

(b) **(i)** Use a pair of compasses and a ruler to bisect each interior angle.

(ii) Explain why these lines are the lines of symmetry of the triangle.

(iii) Where is the centre of rotational symmetry of the triangle?

(c) Cut out the triangle to use as a template.

(d) Use your template to produce a star with rotational symmetry order 12. What is the smallest angle of rotation between triangles?

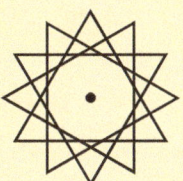

3 A symmetrical 24-pointed star can be made by rotating a regular hexagon.

(a) What is the angle of rotation between hexagons?

(b) Find the other ways in which a regular polygon can be rotated to produce a symmetrical 24-pointed star.

(c) Are any of the stars produced in different ways congruent with each other?

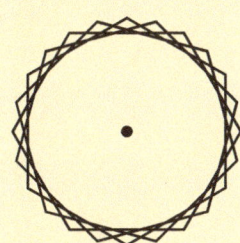

4 Some of these transformations are not possible.

For each one
  (i) say if it is possible
  (ii) explain how it can be done, or why it is not possible.

Each shape is rotated about its centre.

(a) An equilateral triangle producing a 9-pointed star.

(b) An equilateral triangle producing a 12-pointed star.

(c) An equilateral triangle producing an 8-pointed star.

(d) A square producing a 52-pointed star.

(e) A regular nonagon (9-sided polygon) producing a 25-pointed star.

(f) A regular pentagon producing a 25-pointed star.

5 Which of the following stars cannot be made by rotating *any* regular polygon about its centre?

8-pointed, 10-pointed, 11-pointed, 15-pointed, 20-pointed, 17-pointed, 5-pointed

What do the stars you chose have in common?

Finishing off

Now that you have finished this chapter you should be able to:

- describe fully all the symmetries in a shape
- complete part shapes using lines of symmetry
- find the centre of rotational symmetry
- create symmetrical shapes by rotating polygons.

Review exercise

1 **(a)** Copy these patterns and add all the lines of symmetry to each one.

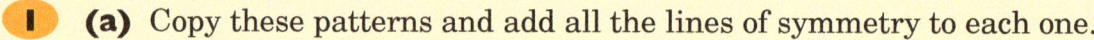

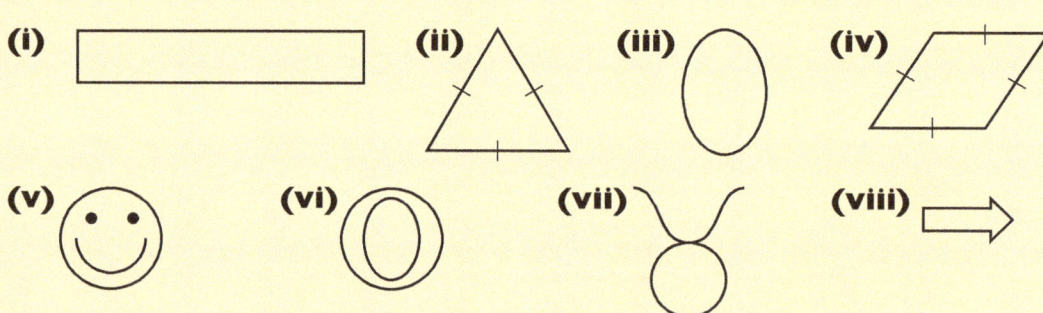

(b) Write down the order of rotational symmetry of each shape.

2 Copy each of the mathematical symbols below.

For each one   **(i)**   add all the lines of symmetry
                   (ii) write down its order of rotational symmetry.

3 The blue patterns are parts of shapes.
The red lines are lines of symmetry of the complete shapes.
Copy the diagrams and show the complete shapes.

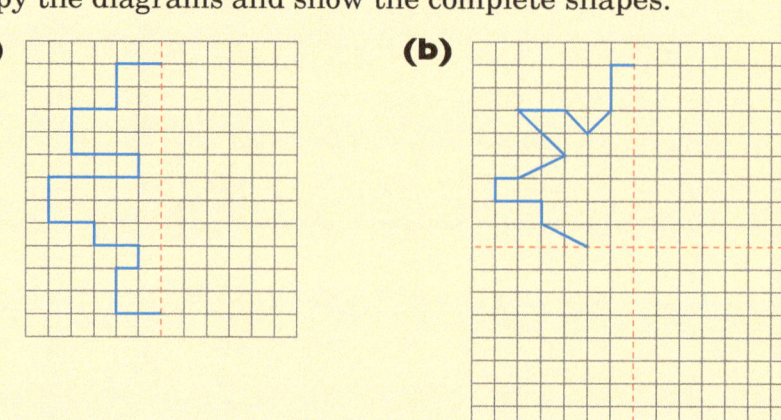

Activity

Snow is made when water vapour freezes to form crystals.
The crystals are usually shaped as regular hexagons.
They combine to form snowflakes with rotational symmetry order 6.

The best known snow crystal photographs
were taken by Wilson Bentley (1865–1931),
a farmer from Vermont, USA.

As a teenager Bentley used photography
and a microscope to capture snow crystal
images on film.

He found no two identical shapes in
over 5300 photographs!

Here are three of his photographs.

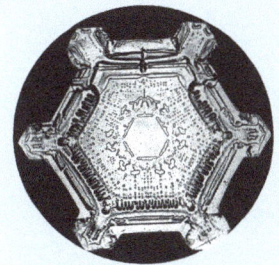  

Here is a snowflake on half-centimetre isometric paper.

Design five more snowflakes.
Make them as large or as small as you like.
They can be very intricate, but remember that each one
must have rotational symmetry order 6.

Put all your class's snowflakes together in a display.
Are any two identical?

Activity   Here are four symmetrical masks from different countries.

Design and make your own full-sized symmetrical mask.

Decide which are the most beautiful, the most ugly and the most
frightening.

20 Equations

Solving equations

Mark and Debbie are playing the Think Of A Number game.

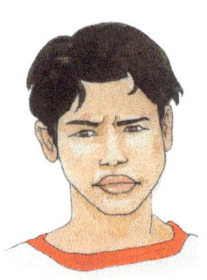

I think of a number.
Multiply it by 4.
Subtract 6.
The answer is 22.

 What is the inverse operation of subtract 6?
What is the inverse operation of multiply by 4?

Mark writes

This is the same as n = 7.

Add 6 to both sides
Tidy up
Divide both sides by 4

$$4n - 6 = 22$$
$$4n - 6 + 6 = 22 + 6$$
$$4n = 28$$
$$4n \div 4 = 28 \div 4$$
$$n = 7$$

 Does it matter whether Mark first adds 6 or divides by 4? Why?
To check Mark's answer work out $4 \times 7 - 6$. How does this help?

Example   Debbie wants to solve the equation $20 - 2x = 6$.

She writes

This is the same as x = 7.

$$20 - 2x = 6$$
Add 2x to both sides $20 - 2x + 2x = 6 + 2x$
Tidy up $20 = 6 + 2x$
Subtract 6 from both sides $20 - 6 = 6 + 2x - 6$
Tidy up $14 = 2x$
Divide both sides by 2 $7 = x$

 How does Debbie deal with the minus sign in $-2x$ in this question?

Task

Play your own game of Think of A Number with a partner.
Make sure you work in the same way as Mark and Debbie.

Jenny and her family go to Avonford Theme Park for the day.
2 adults and 4 children cost €60.   Adults cost €14 each.
Jenny writes the equation $4c + 28 = 60$

 What does c stand for?
How does this equation relate to Jenny's day out?
Solve Jenny's equation.
How much does it cost for one child to go to Avonford Theme Park?

 How would you solve the equation $\frac{x}{4} = 6$?

Exercise

In this exercise set your work out as Mark and Debbie did.

1 Solve the following equations.

    (a) $p + 7 = 13$ **(b)** $10 + f = 14$ **(c)** $h - 5 = 6$ **(d)** $b + 3 = 8$

    (e) $g - 8 = 12$ **(f)** $12 + k = 13$ **(g)** $4a = 16$ **(h)** $3s = 18$

    (i) $5r = 20$ **(j)** $7n = 42$ **(k)** $\frac{c}{3} = 5$ **(l)** $\frac{d}{2} = 3$

    (m) $\frac{g}{5} = 5$ **(n)** $\frac{t}{9} = 4$ **(o)** $10 - f = 4$ **(p)** $12 - k = 3$

    (q) $6 - m = 2$ **(r)** $100 - r = 36$ **(s)** $4x = 80$ **(t)** $\frac{y}{7} = 1$

2 Solve the following equations.

    (a) $5x + 6 = 16$ **(b)** $3a - 7 = 14$ **(c)** $10 + 3f = 19$

    (d) $12 + 7g = 33$ **(e)** $5p - 12 = 8$ **(f)** $11j - 11 = 22$

    (g) $10 - 2n = 4$ **(h)** $15 - 3b = 3$ **(i)** $18 - 4k = 2$

3 Jenny thinks of a number.

I think of a number.
Multiply it by 3.
Subtract 8.
The answer is 19.

    (a) Write an equation to help you find Jenny's mystery number.

    (b) Solve your equation.

    (c) Check you have solved the equation correctly.

4 Petra and her family are going to the cinema. Part of the sign is missing.
Petra pays €25 for 2 adults and 5 children.

    (a) Write down an equation showing this information. Use c for the cost of 1 child.

    (b) Solve your equation to find the cost of one child.

    (c) Check you have solved the equation correctly.

Avonford Cinema

`14.35` Performance

Drinks & Popcorn available in the Foyer at all times

ADULTS €5
CHILDREN €

Investigation You are given that $\frac{12}{x} = 4$. What is the value of x?

Write out the solution to this equation step by step (in the same way as Mark and Debbie on the opposite page).

Equations with an unknown on both sides

Kate wants to buy CDs from her
local shop with her birthday money.

I can buy 4 CDs and
have €2 change or
3 CDs and
have €9 change.

Ben wants to know the cost of one CD.

 You can write this as $4c + 2 = 3c + 9$.
Explain this equation.

Ben writes

$$4c + 2 = 3c + 9$$

Subtract $3c$ from both sides $\quad 4c + 2 - 3c = 3c + 9 - 3c$

Tidy up $\quad\quad\quad\quad\quad\quad\quad c + 2 = 9$

Subtract 2 from both sides $\quad c + 2 - 2 = 9 - 2$

$$c = 7$$

$3c - 3c = 0$

So 1 CD
costs €7.

To check $c = 7$

$$4c + 2 = 3c + 9$$
$$4 \times 7 + 2 = 3 \times 7 + 9$$
$$28 + 2 = 21 + 9$$
$$30 = 30$$

 **How much money was Kate given for her birthday?**

 How would you solve the equation $5n - 3 = 32 - 2n$?

Task

Choose two of the following equations.
Make a poster explaining
(a) how to solve them
(b) how to check your answers.

$$4t - 8 = 22 - t$$

$$4 + 3b = 2b + 6$$

$$4x - 3 = x + 6$$

$$10 - 3p = 12 - 5p$$

$$6 + 3n = 21 - 2n$$

$$r = 20 - 3r$$

 Look at your friends' posters.
Can you follow how they have solved their equations?

Exercise

1 Solve the following equations.
Use your answer to check that you have solved each equation correctly.

(a) $2n + 1 = n + 3$ **(b)** $3p - 2 = 2p + 1$

(c) $1 + 4r = 2r + 5$ **(d)** $3b - 2 = 4 + b$

(e) $10 + 2f = 15 - 3f$ **(f)** $3 + 3w = 10 - 4w$

(g) $5k - 1 = 2k + 2$ **(h)** $15 - x = 23 - 2x$

(i) $12 - 3c = 18 - 5c$ **(j)** $4m - 3 = 5 + 2m$

2 Sarah is buying some bananas from her local shop.

I can buy 5 bananas with 10 cent change or 3 bananas with 46 cent change.

(a) Write down an expression for the cost of 5 bananas with 10 cent change. Use b for the cost of one banana in cent.

(b) Write down an expression for the cost of 3 bananas with 46 cent change.

(c) Make an equation using your two expressions.

(d) Solve your equation to find the cost of one banana.

(e) Check your answer.

(f) How much money does Sarah have?

3 Tim and Brad are having an argument about the following equation.

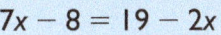

$$7x - 8 = 19 - 2x$$

I think x = 4.

No you are wrong x = 3.

(a) How can you check who is right?

(b) Who has solved the equation correctly?

You do not need to solve the equation.

(c) Show how to solve the equation. Explain your method at each stage.

Solving equations with brackets

Helen and Mike are playing a game of Equation Bingo.
They have a Bingo card each.
The winner is the first one to cross off a complete row or column from their card.

This is Helen's Bingo card.

8	1	4
3	5	9
7	2	6

They draw an equation from a pile.

This is the first equation.

$$5(x + 3) - 2x = 27$$

 **This equation has +, −, × and brackets.
Which do you work out first?**

Remember
BIDMAS

Brackets
Index
Divide
Multiply
Add
Subtract

Helen writes

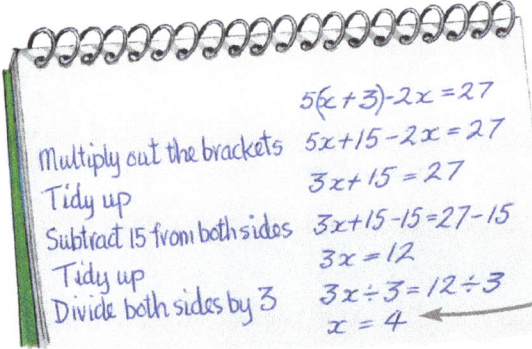

$5(x + 3) - 2x = 27$
Multiply out the brackets $5x + 15 - 2x = 27$
Tidy up $3x + 15 = 27$
Subtract 15 from both sides $3x + 15 - 15 = 27 - 15$
Tidy up $3x = 12$
Divide both sides by 3 $3x \div 3 = 12 \div 3$
$x = 4$

Both Helen and Mike
cross 4 off their Bingo cards.

To check $x = 4$

$$
\begin{aligned}
5(x + 3) - 2x &= 27 \\
5(4 + 3) - 2 \times 4 &= 27 \\
5 \times 7 - 8 &= 27 \\
35 - 8 &= 27 \\
27 &= 27
\end{aligned}
$$

The next equation is $4(p - 2) = 2p + 10$

 How do you solve this equation?

 Helen wins next time. Write down a suitable equation.

Task

Play your own game of Equation Bingo.

 How can you be sure that no one cheats?

Exercise

Set your work out carefully.

1 **(a)** Solve the following equations.

    (i) $4(d + 2) = 20$ **(ii)** $5(h - 3) = 15$

    (iii) $3(2m - 4) = 12$ **(iv)** $7(b + 2) = 21$

    (v) $4(3x - 2) = 16$ **(vi)** $2(5a - 3) = 44$

    (vii) $2(c + 1) - 4 = 0$ **(viii)** $3(y + 2) + y = 26$

    (ix) $5(n - 1) + n = 7$ **(x)** $3(p + 2) + 2(p - 1) = 14$

  (b) Check your answers.

2 Solve the following equations.

  (a) $2(b + 3) = 7b + 1$ **(b)** $5(f - 1) = 4f + 1$

  (c) $4(2e - 3) = 6e + 2$ **(d)** $4(2a + 1) = 5a + 7$

  (e) $3(3g - 8) = 2g - 3$ **(f)** $5(3n - 2) = 10n$

3 **(a)** Find an expression for the perimeter of this rectangle.

  (b) The perimeter of the rectangle is 32 cm. Write down an equation for this information.

  (c) Solve your equation to find the value of b.

  (d) What are the length and width of the rectangle?

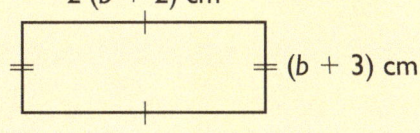

2 (b + 2) cm

(b + 3) cm

Activity Maria and David play a game of Equation Bingo.

Here is   **Maria's card**   **David's card**

1	8	10
9	13	4
5	6	3

13	4	6
12	5	9
1	10	8

To work out who wins the game, solve the following equations.
Check on their cards for the number each time you solve the equation.

1 $5(x - 2) + 3x = 14$ **2** $9c + 7 - 3c = 19$

3 $3 + 4p - 2p = 13$ **4** $10r - 4 = 3(r + 1)$

5 $5(n + 1) = 40$ **6** $16 - 2k = 2$

7 $3(2b - 5) = 4b - 3$ **8** $2(4 - 2d) + 8d = 6d$

Using equations to solve problems

Michelle works at Avonford Rescue Centre.
She feeds the dogs.
Every day she gives each small dog b biscuits.
A large dog gets 2 more biscuits than a small dog.

 Write down an expression for the number of biscuits each large dog receives.

There are 6 small dogs and 4 large dogs at the centre.
They get 38 biscuits altogether.

You can write $6b + 4(b + 2) = 38$.

 Explain this equation.
How many biscuits does each small dog get? What about a large dog?

 **Task**

Here are some expressions

 $A = x - 4$ $E = 3x$ $I = 4x$ $L = x + 6$ $P = x - 1$ $S = 2x$ $W = 3 - x$ $Z = x + 4$

Write down a letter for each of the following. What is the name?

6 more than x   4 less than x   Double x   Twice x   4 times x   3 times x

☐   ☐   ☐   ☐   ☐   ☐

 **Some of the expressions do not have matching words.**
Which are they?
How can you say them in words?

 Task

Graeme, Simon and Tina are given €85 between them.
Graeme is given €m.

(a) Write down expressions for
   (i) Simon's money
   (ii) Tina's money
   (iii) the total for the 3 friends.

(b) Write down an equation.
   Solve it to find m.

(c) How much is each of them given?

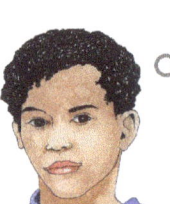

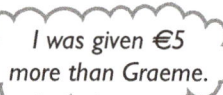

I was given €5 more than Graeme.

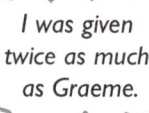

I was given twice as much as Graeme.

 **How can you check your answers?**

Exercise

1 Gemma wants to find the weight of
some identical parcels.
She uses a pair of weighing scales.

She writes

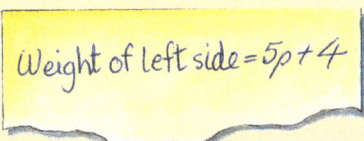

Weight of left side = 5p + 4

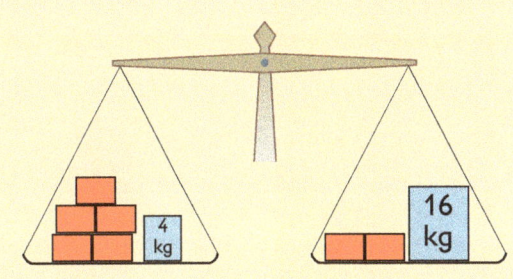

(a) What does *p* stand for?

(b) Write down an expression for the weight of the right-hand side.

(c) Use your expressions and Gemma's expression to make an equation.

(d) Solve your equation to find the weight of one parcel.
Check your answer.

2 John thinks of a mystery number.

*I think of a mystery number.
I multiply my mystery number
by 4 and add 2. The answer is
twice my mystery number plus 4.*

(a) Write down an expression for

(i) '... multiply my mystery number by 4 and add 2'

(ii) '... twice my mystery number plus 4'

(b) Form an equation with your two expressions by making them equal.

(c) Solve your equation to find John's mystery number.

3 Peter is the penguin keeper at Avonford Zoo.
He is designing a new enclosure for them.
Here are his two designs.

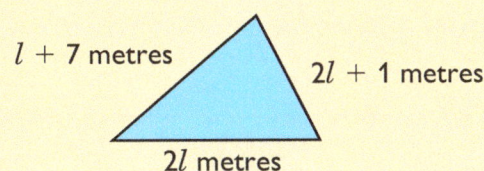

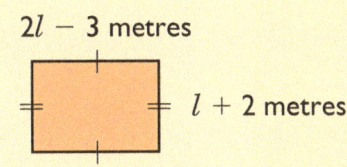

$2l - 3$ metres

$l + 7$ metres

$2l + 1$ metres

$l + 2$ metres

$2l$ metres

(a) Write down expressions for the perimeters of the enclosures. Simplify
your answers.

Both penguin enclosures have the same perimeter of fencing.

(b) Write down an equation for *l*.

(c) Solve your equation to find *l*.

(d) How much fencing does Peter need to build one enclosure?

(e) What are the dimensions of each penguin enclosure?

Finishing off

> **Now that you have finished this chapter you should be able to:**
>
> - understand the difference between an equation and an expression
> - solve an equation with an unknown on one side
> - solve an equation with an unknown on both sides
> - use equations to solve problems.

Review exercise

1 Solve the following equations.

  (a) $p + 5 = 9$ **(b)** $t - 3 = 11$

  (c) $10 + a = 12$ **(d)** $2 + r = 7$

  (e) $2y = 12$ **(f)** $4m = 18$

  (g) $\dfrac{k}{3} = 6$ **(h)** $\dfrac{b}{7} = 9$

2 Solve the following equations.

  (a) $4n + 7 = 15$ **(b)** $4 + 3d = 19$

  (c) $2f - 7 = 3$ **(d)** $6h - 5 = 1$

  (e) $12 - 3t = 3$ **(f)** $24 - 7p = 10$

  (g) $4(k + 2) = 24$ **(h)** $7(6 - 2r) = 28$

3 Solve the following equations.

  (a) $5t + 8 = 4t + 18$ **(b)** $3n - 5 = 7 - 3n$

  (c) $4(g + 1) = 3g + 9$ **(d)** $2(p - 2) = p - 1$

  (e) $3(x + 2) = 2x + 13$ **(f)** $10 - 3r = 12 - 5r$

  (g) $3(2h - 3) = 4h + 3$ **(h)** $2(5 - k) = 12 - 4k$

4 Tim is buying some fish and chips.
He is charged €12 for 3 portions of
fish and 3 portions of chips.

  (a) Write down an equation to show
this information.
Use f to represent the cost of one
portion of fish.

  (b) Solve your equation to work out the price of one portion of fish.

  (c) Check your answer.

5 Karl is buying some pencils from the local shop.

> I can buy 6 pencils with 16 cent change or 4 pencils with 44 cent change.

(a) Write down an expression for the cost of
   (i) buying 6 pencils and having 16 cent change
   (use p for the cost in cent of one pencil).
   (ii) buying 4 pencils and having 44 cent change.

(b) Form an equation with your two expressions by making them equal.

(c) Solve your equation to find the cost of one pencil. Check you have solved the equation correctly.

(d) How much money does Karl have?

6 Caroline is weighing some soup using scales. The soup tins are all the same. Use m kilograms for the mass of one soup tin.

She writes

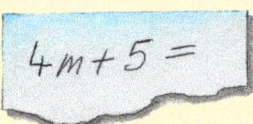

$$4m + 5 =$$

(a) How should Caroline finish off this equation?

(b) Solve the equation to find the weight of one tin of soup.

(c) Check you have solved the equation correctly.

(d) What is the total weight on the left-hand side of the scales?

7 Oliver and Susan are trying to solve this equation.

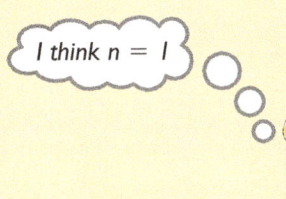

> I think n = 1

$$2(7 - 3n) = 10n - 18$$

> No n = 2

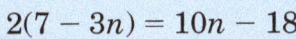

(a) Check who is right *without* solving the equation.

(b) Show clearly how to solve the equation.

Activity   Design your own set of Equation Bingo cards.

21 Probability

Using probability

'There's a 10% chance of showers tomorrow'

'There is a 30% chance a car develops a fault in its first year'

'The chance of winning the lottery is 1 in 14 million'

'England has a 50 : 50 chance of winning this match'

? **How do you think each person decided their probability?**

Calculating probabilities using equally likely outcomes

When you throw a die there are six possible numbers, or **outcomes**.
Each outcome is *equally likely* to come up.

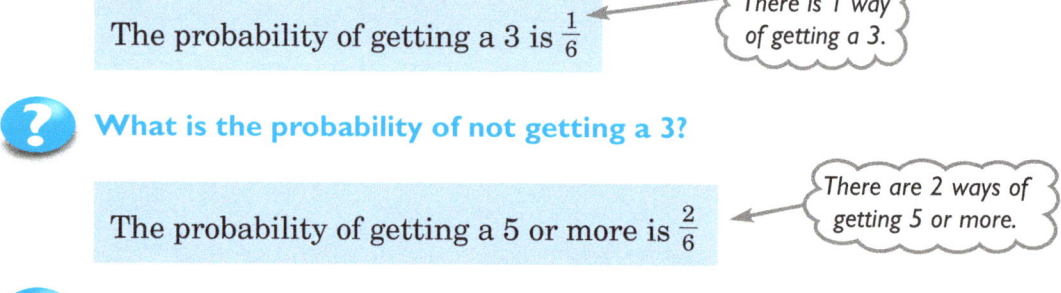

The probability of getting a 3 is $\frac{1}{6}$

There is 1 way of getting a 3.

? **What is the probability of not getting a 3?**

The probability of getting a 5 or more is $\frac{2}{6}$

There are 2 ways of getting 5 or more.

? **What is the probability of not getting 5 or more?**

Task

1 Think of six different events whose probabilities can be found using equally likely outcomes.

2 Write down each event, its probability, and the probability that the event does not happen.

? **What do you notice about the relationship between the probability of an event and the probability that it does not happen?**

? **Look at the probabilities in the examples at the top of the page. Which can you find using equally likely outcomes?**

Exercise

1 Sunil is using this 8-sided spinner in a game.
What is the probability that he gets

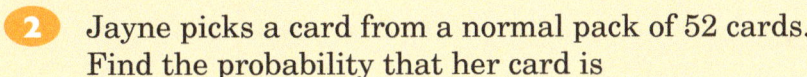

   (a) yellow   **(b)** red

   (c) 1   **(d)** a green 4

   (e) a red or yellow number 3 or 4   **(f)** a yellow 2?

2 Jayne picks a card from a normal pack of 52 cards.
Find the probability that her card is

   (a) the seven of diamonds   **(b)** an ace   **(c)** not an ace

   (d) a spade   **(e)** not a spade   **(f)** a king, queen or jack.

3 In the British National Lottery, balls are selected at random by a machine.
They are numbered from 1 to 49.
What is the probability that the first ball is

   (a) the number 42   **(b)** odd   **(c)** greater than 30

   (d) a multiple of 3   **(e)** not a multiple of 3   **(f)** a prime number

   (g) not a prime number   **(h)** an even number greater than 20?

4 James throws a 10-sided die numbered 1 to 10.

   (a) What is the probability he gets a 6?

   (b) What is the probability he gets a number less than 6?

   (c) What is the probability he gets a number greater than 6?

   (d) **(i)** Add up these three probabilities.
       (ii) Explain your answer.

5 A local newspaper states that the probability that Avonford Town will win
their next match is $\frac{3}{10}$, and the probability they will lose is $\frac{6}{10}$.

What is the probability that the match will be a draw?

6 Zoë and Anna are playing a game with cards numbered from 1 to 12.
Anna takes a card.

   (a) Zoë says 'You win if you get a prime number, otherwise I win.'
       Is Zoë's game fair? Explain your answer.

   (b) Anna says 'I win if I get a prime number, you win if you get a multiple
       of 4, otherwise it's a draw.'
       Is Anna's game fair? Explain your answer.

   (c) Think of a way to make the game fair.

Estimating probability

In the game Pass the Pigs two small model pigs are thrown in the air. You score points according to how they land.

Unlike a die, the different positions for the pigs are not equally likely. However, you can **estimate** the probabilities by carrying out an experiment.

Kerry threw some of the pigs 50 times. Here are her results.

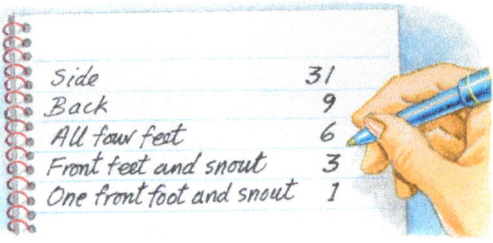

Side	31
Back	9
All four feet	6
Front feet and snout	3
One front foot and snout	1

Relative frequency is used as an estimate of probability.

From these results, the **relative frequency** of a pig landing on its side is $\frac{31}{50}$, or 0.62. Each time Kerry throws the pig, there is a probability of 0.62 that it lands on its side.

Task

To make a biased die
- **(a)** cut out the net of an ordinary die
- **(b)** attach a small piece of blu-tack to the inside of one face
- **(c)** stick the net together.

Throw your die 30 times and write down your results.

Use your results to estimate the probability of getting each number.

 How can you make your estimates more reliable?

Subjective estimates

There is a 10% chance of rain tomorrow.

There is a chance in a million of finding life on another planet.

 Steve and Sue could not have worked out their probabilities using equally likely outcomes. Could they perform experiments to estimate their probabilities?

These statements are called **subjective estimates** of probability.

 How do you think Steve and Sue decided on the estimates of their probabilities? How accurate are they?

Exercise

1 Paul is carrying out an opinion poll
for a local election.
He asks 80 people which candidate
they intend to vote for.
Here are his results.

Joe Wilson 32
Mary Smythe 38
Fred Jones 10

 (a) Estimate the probability that the next person he asks will vote for is
  (i) Joe Wilson   **(ii)** Mary Smythe.

 (b) 2000 people vote in the local election.
  From Paul's opinion poll, how many people would you expect to
  vote for each candidate?

2 Martin thinks his die is biased. He thinks
the number six comes up more often than
the others.
He throws the die 120 times.
Here are his results.

Score	Frequency
1	20
2	16
3	19
4	14
5	15
6	36

 (a) How many times would you expect
  each number to come up if the die is fair?

 (b) From Martin's results, estimate the
  probability of getting a six on his die.

 (c) Do you think the die is fair?

3 Gemma, Simon and Julie have three
coins. They know that one of the
coins is biased. They each toss one
of the coins a different number of times.

	Gemma	Simon	Julie
Heads	21	61	90
Tails	29	39	110

 Their results are shown opposite.

 (a) How many times does each of
  them toss their coin?

 (b) Whose results are the most reliable?

 (c) Which coin do you think is biased? Can you be sure?

4 **(a)** Estimate the probability, as a percentage, of each of the following events.
  (i) You will have three or more children.
  (ii) You will use a computer some time in the next week.
  (iii) Mankind will discover time travel in the next hundred years.
  (iv) It will rain tomorrow.

 (b) Draw a probability scale and mark each event on it.

5 Write down

 (a) three situations where the probability of the outcomes can be
  calculated using equally likely outcomes

 (b) three situations where the probabilities can be estimated using an
  experiment

 (c) three situations where the probabilities must be subjective estimates.

Two or more events

Carly and Richard are playing games with coins.

Let's take turns to toss two coins. I win with two heads or two tails, and you win with one of each.

That's not fair! You've got two chances of winning and I've only got one.

? **Do you think the game is fair?**

Task

1 Play Carly and Richard's game 20 times.

2 Make a tally chart to show the results.

3 Compare your results with the rest of the class.

Carly and Richard use a 20 cent coin and a 50 cent coin.

They make a table to show all the possible ways the coins can land.

	20 cent coin	
50 cent coin	**Heads** (H)	**Tails** (T)
Heads (H)	H H	H T
Tails (T)	T H	T T

? **What is the probability of getting**
(a) two heads
(b) two tails
(c) one head and one tail?

? **What is the probability that Richard wins?**

Exercise

1 Marie is deciding what to
have for lunch.
She wants two courses.

(a) Make a list of all the
possible lunches for Marie.

(b) How do you work out the
number of possible lunches?

Lunch Menu

First course	Second course
Cod and chips	Apple pie
Lasagne	Cheesecake
Cheese salad	Trifle
Chicken curry	

2 Some children are playing Snakes and Ladders.
They use two dice.
Copy and complete this table
to show all the possible outcomes
when the two dice are thrown.

(a) James has the next turn and
needs 5 to win.
What is the probability he
wins on this turn?

	1	2	3	4	5	6
1	2	3	4			
2	3					
3						
4						
5						
6						12

(b) Mark will land on a snake if he gets 10.
What is the probability he lands on a snake?

(c) Fiona will land on a ladder if she gets 4 or 11.
What is the probability she lands on a ladder?

3 A spinner has 5 sides, numbered 0 to 4.
Jessica spins it twice and adds up her two scores.

(a) Make a list or table to show all the possible outcomes.

(b) What is the probability that Jessica gets

(i) 3   **(ii)** 4   **(iii)** more than 5   **(iv)** both scores the same?

4 In a game, two dice are used. One has 4 sides, labelled 2, 4, 6 and 8.
The other has 8 sides, labelled 1 to 8.
The score is the difference between the numbers on each die.

(a) Make a table or list showing all the possible outcomes.

(b) Find the probability of scoring

(i) 0   **(ii)** 1   **(iii)** 2   **(iv)** 5 or more.

Activity James and Kate decide to play a game with three coins.
James wins if they get three heads or three tails,
and Kate wins otherwise.

(a) Make a list of all the possible outcomes of
tossing three coins.

(b) What is the probability James wins?

(c) What is the probability Kate wins?

(d) Think of a different set of rules to make the game fair.

Finishing off

Now that you have finished this chapter you should be able to:

- find the probability of an event using equally likely outcomes
- know that the probabilities of an event happening or not happening add up to 1
- know that the probabilities of all possible outcomes add up to 1
- use relative frequency to estimate probabilities
- make subjective estimates of probability
- find all the possible outcomes of two events.

Review exercise

1 Aisha throws an 8-sided die, numbered 1 to 8.
Find the probability that the number she gets is

(a) 1   **(b)** not 1

(c) 6 or more   **(d)** an odd number

(e) a square number   **(f)** not a square number.

2 Emma and Rachel are playing a game of 'Battleships'.
Emma has drawn this diagram. The red squares represent ships.

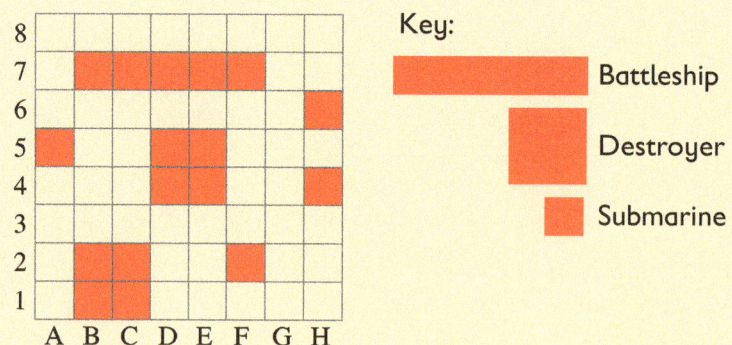

Without looking at the diagram Rachel chooses a square at random.
Find the probability she

(a) hits one of Emma's ships

(b) does not hit a ship

(c) hits the battleship

(d) hits a destroyer

(e) hits a submarine.

(f) Add up your answers to (b), (c), (d) and (e). What do you notice?

3 Dave lives on a main road.
One morning he carries out a survey of the vehicles passing his house between 10 am and 11 am.
Here are his results.

Car	Lorry	Motorcycle	Van	Bus/Coach
43	18	11	16	7

(a) How many vehicles are in Dave's survey?

(b) Estimate the probability that the next vehicle that passes Dave's house
    (i) after 11 am is a car
    (ii) after 5 pm is a lorry.

(c) How reliable are your answers to (b)(i) and (ii)? Explain.

4 For the following events

    (i) estimate the probability
    (ii) explain how you chose each probability
    (iii) show your results on a probability scale.

(a) Mankind will land on the moon again in the next 10 years

(b) You will live to be 100 years old

(c) You will eat chips sometime in the next week

(d) You will watch TV for more than one hour tonight

(e) It will snow sometime next January.

5 Jamie throws two four-sided dice. Each die is numbered 0, 1, 2 and 3.
He multiplies the two numbers together to get his score.

(a) Make a table to show all the possible outcomes.

(b) What is the probability that Jamie scores 0?

(c) What is the probability he scores 6?

(d) What is the probability his score is a square number?

6 Lee and Kirsty are playing a game.
They have 6 playing cards: the king, queen and jack of spades and the king, queen and jack of hearts.
They take turns to pick two cards at random.

(a) Make a list of all the possible outcomes of picking two cards.

What is the probability of getting

(b) two spades

(c) a heart and a spade

(d) two kings

(e) a queen and a jack?

Working systematically

Look at the diagram.
You can move from the red circle to any of the other circles.
You can only move downwards.

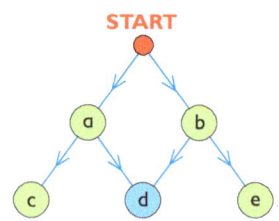

? **In how many different ways can you move from the red circle to each of the green circles?**
What about the blue circle?

Task

1 Work out how many different ways there are of moving from the start to each of the green circles.
An example is done for you.

2 Describe any patterns you notice.

Now look at the row of blue circles at the bottom of the diagram.

3 How many ways are there of reaching each of these circles?

4 Find the total of each row in the triangle.
What do you notice?
What is special about these numbers?

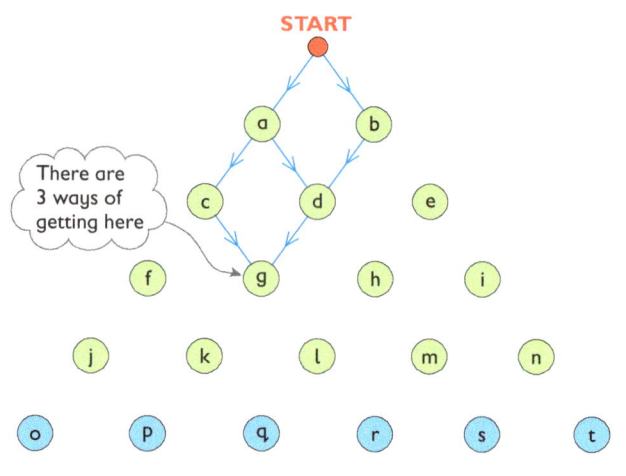

There are 3 ways of getting here

? **You know the numbers in one row.**
How can you use them to find the numbers in the next row?

? **It helps to be systematic.**
How can you be systematic solving this problem?
What are the advantages of working systematically?

This pattern of numbers is called Pascal's Triangle. Pascal's Triangle has many different number patterns.

? **Look at the diagonal lines. What patterns can you find?**

Investigation

The Tower of Hanoi is a
mathematical puzzle.
There are three pegs and several
discs as shown.

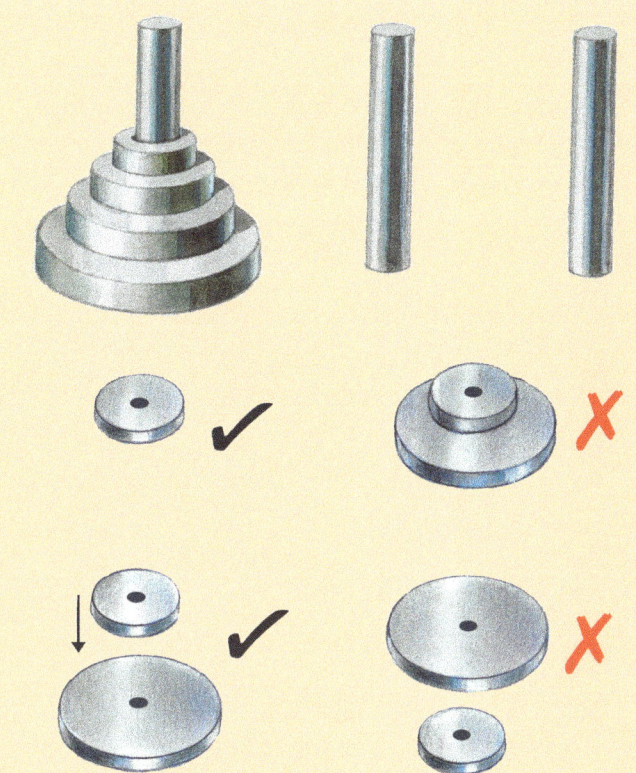

The aim is to move all of the discs
onto another peg.
The rules are:

> You can only move one disc
> at a time.
>
> You must not put a larger disc
> on top of a smaller disc.

*You should take as
few moves as possible.*

1 Make your own Tower of Hanoi by cutting out discs from card.

2 Start with 2 discs. How many moves do you need?

3 Show that it cannot be done in fewer moves.

4 How many moves are needed with

  (a) 3 discs   **(b)** 4 discs   **(c)** 10 discs   **(d)** 15 discs?

5 Explain how you have done this investigation systematically.
  How has it helped you?

Investigation

Look at this addition sum.

There are 10 letters. Each letter stands for one
(and only one) of the 10 numbers
1, 2, 3, 4, 5, 6, 7, 8, 9, 0.

```
  W R M G M T Y
  H M H T M T K
  ─────────────
  M I K T O G K Y
```

1 Replace all the letters by their numbers.
  Complete this table.

Number	1	2	3	4	5	6	7	8	9	0
Letter										

2 You can solve this problem systematically.
  Write down how you do this, step by step.

Making predictions

Maria is laying a path around her garden pond.
She uses regular pentagons as paving stones.
Each paving stone has sides of length 1 foot.

Here are the first three stages of her design:

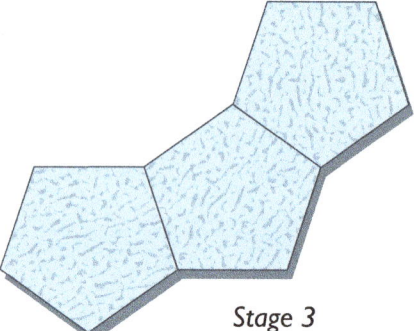

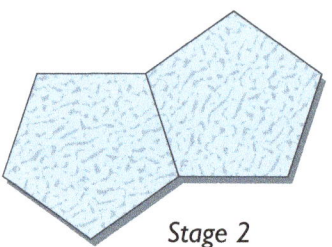

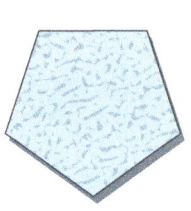

Stage 3

Stage 2

Stage 1

She makes a table giving the
perimeter of her path at
different stages.

Number of paving stones	Perimeter (in feet)
Stage 1	5
Stage 2	8
Stage 3	11

 What is the perimeter for each of the next three stages?
How can you check your answers?

Maria writes

Each stage has a perimeter of 3 more than the one before.

So perimeter of 7th stage = 5 + 6 x 3 = 23.

 Why does this calculation work?
How many paving stones does Maria need to complete her path?

 What calculation should you do to find the perimeter of
(i) the 5th stage (ii) the 8th stage?
What about the final stage? Can you still call it a perimeter?

Task

1 Draw the first three stages of paths made out of regular hexagonal paving stones.
2 Is it possible to use hexagonal paving stones to make a path around a pond
  without leaving any gaps?

Each stone has sides of 1 metre.
3 Find the perimeter of each of the stages you have drawn.
4 Predict the perimeter of the next stage.
5 Show how you would work out the perimeter of the 5th and the last stage.

Exercise

It will help if you cut out templates to draw round, for both questions.

1 Avonford High School's canteen uses trapezium-shaped tables.
Each table can seat 7 people.
The tables are laid in long lines.

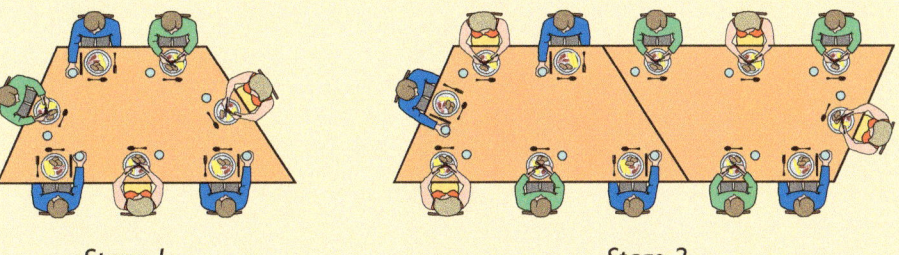

Stage 1   *Stage 2*

(a) Draw the next two stages.

(b) Make a list showing how many people can be seated at stage 1, 2, 3 and 4.

(c) Predict how many people can be seated at stages 5, 6 and 7.

(d) How many people could be seated at a line of 20 tables?

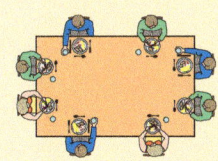

(e) Avonford High School also use tables like these:
Repeat parts (a)–(d) for these tables.

2 Rob is laying a path using regular octagonal paving stones.
Each side is one foot long.

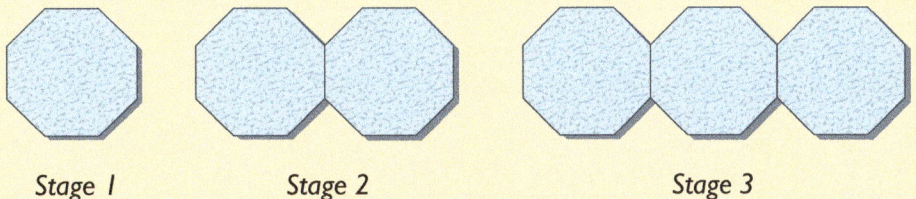

Stage 1   *Stage 2*   *Stage 3*

(a) Draw the next three stages.

(b) Make a table showing the perimeter of the first five stages.

(c) What do you need to add on to find the perimeter of the next stage?

(d) Predict the perimeter of the next three stages.

(e) Show how you would work out the perimeter of the 20th and the 50th stage.

Investigation   Rob uses 8 octagonal paving stones to lay a path around a pond.
Draw this in a diagram. There are three possible designs.

What is the perimeter of the path?

Finding the rule

Sam is making a tiling pattern on his bathroom floor.
It has layers of different tiles.

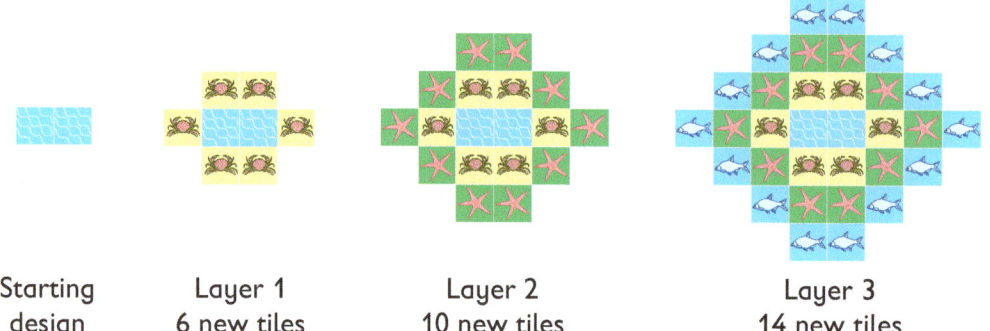

| Starting design | Layer 1 6 new tiles | Layer 2 10 new tiles | Layer 3 14 new tiles |

Sam wants to know how many new tiles he will need for the 10th layer.
He makes a table to help him find a rule.

Number of layers, L	Number of new tiles, T	What I see: my rule
1	6	1 × 4 + 2
2	10	2 × 4 + 2
3	14	3 × 4 + 2
4	18	
5	22	

? **Why does Sam use the letters *L* and *T*?**

? **Copy and complete Sam's new table.**

Sam notices a pattern. The number of new tiles increases by 4 for each new layer.
Sam says 'It's the 4 times table; add 2'.

? **You can write Sam's rule as the formula $T = 4L + 2$. Explain this rule.**

Task

1 How many tiles are needed for each of the first 5 layers of these starting designs?

  (a) ☐   **(b)** ☐☐☐

2 Find a rule for the number of tiles needed for each layer in the same way as Sam did.

Exercise

1 Draw the next three layers for these tiling designs.

(i)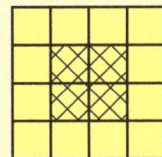

(ii)

For each design

(a) find how many new tiles are needed for each layer

(b) make a table of your results

(c) **(i)** find how much the number of new tiles is increasing by for each new layer

(ii) find the rule.

2 Jenny has investigated some other tiling patterns. Here are her results.

(a) Complete Jenny's tables of results.

(b) Find rules for the number of tiles needed.

Layer	Number of new tiles	Layer	Number of new tiles
1	7	1	3
2	12	2	9
3	17	3	15
4	22	4	21
5	27	5	27
6	?	6	?
10	?	50	?

Investigation Investigate your own tiling patterns. Start by looking at the following:

Is there a rule for every starting design?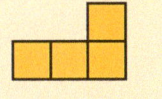

Make a poster of your favourite designs. On your poster explain how you found the rule in each case.

Investigation

Here is part of a spreadsheet.

The numbers in the columns B to G are formed from column A, using these formulae:

= SUM(A) + 6   = SUM(A) * 6 + 1

= SUM(A) * 7   = SUM(A) * 3 + 4

= SUM(A) * 4 + 3   = SUM(9 − (A) * 2)

	A	B	C	D	E	F	G
1	1	7	7	7	7	7	7
2	2	11	14	13	5	10	8
3	3	15	21	19	3	13	9
4	4	19	28	25	1	16	10
5	5	23	35	31	−1	19	11
6	6	27	42	37	−3	22	12

Which formulae is used for which column?

Sequences

Look at these numbers   2, 7, 12, 17, ...
A list of numbers is called a **sequence**.

 What are the next three numbers in the sequence?
What is the rule to find these numbers?

The fourth **term** in this sequence is 17.

> *The **difference** is always 5.*

 What is the sixth term in the sequence?
What is the tenth term?

Kim wants to find the fifth term in the sequence.

She writes

> *To find the 5th term I need to find 2 plus 4 lots of 5.*

$+5 \quad +5 \quad +5 \quad +5$
2 \quad 7 \quad 12 \quad 17 \quad ?
5th term = 2 + 4 × 5
so the 5th term is 22

Kim makes a table for the **formula** $3n + 2$.

> *This can be written as the **function** $n \to 3n + 2$*

Term number	1	2	3	4	5
Term	5	?	?	14	?

> *For example*
> $3 \times 4 + 2 = 14$

 What are the missing ?'s ?
What is the 50th term in the sequence?

Task

Match these sequences to their rules:

Add 3 to the previous term. First term 4.

$n \to 4n - 1$

$n \to 3n - 5$

Add the previous two terms together. First term 1, 1.

$n^2 + 3$

$3 + 4(n - 1)$

Double the previous term. First term 1.

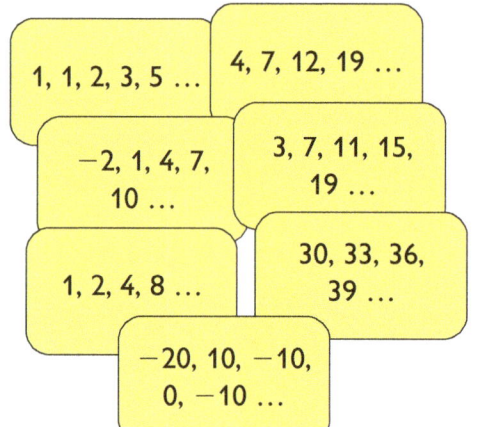

1, 1, 2, 3, 5 ...

4, 7, 12, 19 ...

−2, 1, 4, 7, 10 ...

3, 7, 11, 15, 19 ...

1, 2, 4, 8 ...

30, 33, 36, 39 ...

−20, 10, −10, 0, −10 ...

Exercise

1 Write down the next three terms for each of the following sequences:

(a) 512, 256, 128, 64, ..., ..., ...   **(b)** 2, 5, 8, 11, 14, ..., ..., ...

(c) 40, 36, 32, 28, ..., ..., ...   **(d)** 1, 4, 9, 16, 25, 36, ..., ..., ...

(e) 1, 2, 4, 8, 16, 32, 64, ..., ..., ...   **(f)** $-32, -16, -8, -4, ..., ..., ...$

(g) $\frac{1}{2}, \frac{2}{3}, \frac{3}{4}, \frac{4}{5}, \frac{5}{6}$ ..., ...

For each of the sequences (a)–(g) write down in words the rule you are using.

2 Find the 20th term in each of the following sequences:

(a) 3, 5, 7, 9, ...   **(b)** 4, 7, 10, 13, ...   **(c)** 1, 5, 9, 13, ...

3 Here are formulae for the nth term of some sequences.
Write down the first four terms for each of them.

(a) $5n + 4$ **(b)** $n \to 2n - 1$ **(c)** $n^2 + n$

4 A snail is crawling up a tree. The tree is 21 m tall.

Every day it crawls up 4 m but every night it falls back 2 m.

(a) Write down the first 6 terms in the number sequence that describes the snail's height at the start of each day.

(b) How long does it take for the snail to reach the top of the tree?

5 **(a)** Write down the number of dots in each of these patterns.

(b) How many dots are there in each of the next 5 patterns?

These numbers are called **triangular numbers**.

The formula for the nth triangular number is $\dfrac{n(n + 1)}{2}$.

(c) Check that this formula works for the first four triangular numbers.

(d) Find
    (i) the 10th   **(ii)** the 20th   **(iii)** the 100th
triangular numbers.

Finishing off

> **Now that you have finished this chapter you should be able to:**
>
> - work systematically to solve problems
> - continue patterns
> - generate terms of a sequence given the rule or formula
> - make predictions
> - write down a rule
> - find a formula for the nth term.

Review exercise

1 Alison is making some patterns out of matches.

Pattern 1   Pattern 2   Pattern 3

(a) Draw the next two patterns.

(b) Copy and complete the following table.

Pattern	1	2	3	4	5
Number of matches					

(c) How many matches are being added on each time?

(d) How many matches are needed for
   (i) the 6th pattern   **(ii)** the 10th pattern   **(iii)** the 100th pattern?

(e) Write down a formula for the number of matches needed for any pattern.
   Use P for the pattern number and M for the number of matches required.

(f) Repeat parts (a)–(e) for the following pattern of matches.

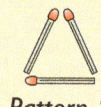   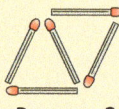   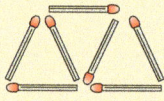

Pattern 1   Pattern 2   Pattern 3

2 Tim investigates two other patterns of matches.
Here are his results:

(i)

Pattern number	Number of matches
1	1
2	5
3	9
4	13

(ii)

Pattern number	Number of matches
1	10
2	13
3	16
4	19

For each table answer the following questions:

(a) How many matches are being added on each time?

(b) How many matches are needed for pattern number
(i) 5   (ii) 10   (iii) 20?
Explain how you reached your answers.

(c) Write down a formula for the number of matches needed for the pth pattern
Use P for the pattern number and M for the number of matches required.

3 Here are the formulae for the nth term of a sequence.

Write down the first five terms for each of them.

(a) $3n + 4$ **(b)** $20 - 2n$ **(c)** $n^2 + 4$

4 Kerry thinks of the following sequence:

2, 6, 10, 14, …

(a) What are the next three terms of Kerry's sequence?

(b) What is the 20th term?

(c) Find a formula for the nth term.

(d) Use your formula to find the 50th term.

Activity   A sequence begins with the numbers 1, 2, …

Find three different ways of continuing this sequence.

For each one, write down a rule for the sequence.

Complete the multiplication $12 \times ? = 4$

That is impossible. When you multiply the answer is bigger not smaller.

It must be a mistake. It should be divide. $12 \div 3 = 4$

 When does a multiplication produce a smaller answer?
Complete the calculation $12 \times ? = 4$

Task

1 From the calculations below select any that have the same answers.

(a) $6 \times \frac{1}{2}$ **(b)** $8 \div 4$ **(c)** $\frac{1}{5}$ of 10   **(d)** $6 \div 2$

(e) $\frac{3}{4}$ of 20   **(f)** $\frac{1}{4}$ of 8   **(g)** $20 \times \frac{3}{4}$ **(h)** $10 \div 5$

(i) $(20 \div 4) \times 3$ **(j)** $\frac{1}{2}$ of 6   **(k)** $10 \times \frac{1}{5}$ **(l)** $8 \times \frac{1}{4}$

2 Write down two or more calculations that give the same answer as

(a) $18 \times \frac{1}{6}$ **(b)** $36 \div 9$ **(c)** $\frac{1}{3}$ of 15   **(d)** $(12 \div 6) \times 5$

Dividing fractions by a whole number

Three days before the end of her holiday Jane realises that she only has $\frac{1}{5}$ of a bottle of suntan lotion left.
What fraction of the bottle can she use each day?
Jane wishes to work out $\frac{1}{5} \div 3$
To do this she works out $\frac{1}{5} \times \frac{1}{3} = \frac{1}{15}$
She can use $\frac{1}{15}$ of the bottle each day.

July

~~18~~ ~~19~~ ~~20~~ ~~21~~ ~~22~~ ~~23~~ ~~24~~
~~25~~ ~~26~~ ~~27~~ 28 29 30 (31)
Arrive home

 Is $\frac{1}{15}$ smaller than $\frac{1}{3}$?

 How can you find a multiplication that gives the same answer as dividing by a whole number?

 Write down a multiplication that will give the same answer as
(a) $\frac{1}{4} \div 3$ **(b)** $\frac{2}{3} \div 2$ **(c)** $\frac{4}{5} \div 8$.
Work out each calculation.

⚠ Remember to cancel whenever possible.

 Explain why the calculation $\frac{2}{3} \times \frac{1}{4} \times \frac{3}{5}$ will help you to work out $\frac{2}{3} \div 4 \times \frac{3}{5}$.
Complete the calculation. Make sure you cancel first.

Exercise

1 To practise multiplying fractions, work out

(a) $\frac{1}{5} \times \frac{1}{3}$ (b) $\frac{1}{2} \times \frac{1}{4}$ (c) $\frac{2}{3} \times \frac{1}{5}$

(d) $\frac{3}{4} \times \frac{2}{3}$ (e) $\frac{1}{2} \times \frac{1}{3} \times \frac{1}{4}$ (f) $\frac{1}{2} \times \frac{2}{3} \times \frac{1}{4}$

(g) $\frac{2}{3} \times \frac{3}{4} \times \frac{4}{5}$ (h) $\frac{5}{8} \times \frac{4}{10} \times \frac{2}{3}$ (i) $\frac{1}{2}$ of $\left(\frac{4}{7} \times \frac{21}{25}\right)$

(j) $\frac{1}{4}$ of $\left(\frac{3}{5} \times \frac{4}{9}\right)$ (k) $\frac{1}{2} \times \frac{2}{3} \times \frac{3}{4} \times \frac{4}{5}$ (l) $\frac{2}{3} \times \frac{3}{4} \times \frac{4}{5} \times \frac{5}{6} \times \frac{6}{7}$

2 Work out each of the following. Then write down two equivalent calculations in the same way as you did for part 2 of the Task opposite.

(a) $\frac{1}{3}$ of 18   (b) $\frac{1}{4} \times 24$ (c) $\frac{2}{3}$ of 12   (d) $(18 \div 6) \times 5$

3 Change each of the following into a multiplication before evaluating.

(a) $\frac{1}{6} \div 3$ (b) $\frac{1}{4} \div 5$ (c) $\frac{2}{5} \div 5$

(d) $\frac{3}{5} \div 3$ (e) $\frac{4}{7} \div 2$ (f) $\frac{6}{7} \div 3$

4 Work out the following.

(a) $\frac{5}{9} \div 5$ (b) $\frac{10}{11} \div 5$ (c) $\frac{1}{2} \div 4 \times \frac{4}{5}$ (d) $\frac{1}{3} \div 2 \times \frac{9}{10}$

(e) $\frac{1}{2}$ of $\left(\frac{1}{3} \div 2\right)$ (f) $\left(\frac{2}{7} \times \frac{3}{4}\right) \div 6$ (g) $\frac{5}{9} \times \frac{3}{4} \div 5$ (h) $\frac{3}{5} \times \frac{15}{16} \times \frac{8}{9} \div 4$

5 Anna wishes to share $\frac{1}{4}$ of her birthday cake between 5 people. What fraction of the cake will each get?

6 George has read $\frac{1}{5}$ of his book in 2 days.

(a) What fraction is this per day?

(b) What fraction has he still not read?

He must return the book to the library in 6 days' time.

(c) What fraction must he now read each day?

7 Martin Miller the Millionaire left his fortune to be divided between his children and his grandchildren.

(a) A quarter of his money is to be divided equally between his three children.
What fraction does each receive?

(b) The remainder is divided equally between his ten grandchildren.
What fraction is this for each grandchild?

(c) His fortune was €4 million.
Calculate the amount that went to each child and each grandchild.

Reciprocals

Each glass holds $\frac{1}{4}$ of a bottle of lemonade.
How many glasses can be filled from 3 bottles?

The problem can be solved as a division

$$3 \div \frac{1}{4} = 12$$

How many $\frac{1}{4}$s in 3?

It can also be solved as a multiplication

$$3 \times 4 = 12$$

One bottle can fill 4 glasses.
How many glasses can
three bottles fill?

 How many glasses can be filled from $\frac{3}{4}$ of a bottle of lemonade?
Explain how the calculations $\frac{3}{4} \div \frac{1}{4}$ and $\frac{3}{4} \times 4$ can be used to answer this.

4 is known as the *reciprocal* of $\frac{1}{4}$.
The fraction $\frac{1}{3}$ is the *reciprocal* of 3.
The answer to a division can be found by *multiplying by the reciprocal*.

 Write down the reciprocals of $\frac{1}{2}$, 7, $\frac{1}{5}$, 6, $\frac{1}{9}$.

Task

Copy and complete the table below.

Division	Multiply by the reciprocal	Check by multiplying
$5 \div \frac{1}{4} =$	$5 \times 4 = 20$	$20 \times \frac{1}{4} = 5$
$4 \div \frac{1}{7} =$		
$\frac{1}{2} \div 3 =$		
$\frac{1}{4} \div \frac{1}{8} =$		
$3 \div \frac{3}{4} =$	$3 \times \frac{4}{3} = 4$	$4 \times \frac{3}{4} = 3$
$8 \div \frac{4}{5} =$		

Look carefully at the last two rows.

 How does the calculation $4 \times \frac{3}{4} = 3$ help you work out $3 \div \frac{3}{4}$?

What is the reciprocal of $\frac{3}{4}$? Write down the reciprocal of $\frac{4}{5}$.

 Look back at all your answers.
When has a division produced a larger answer than the number you started with?
When do you get a smaller answer?

Exercise

1 Write down the reciprocals of

    (a) 4   **(b)** $\frac{1}{2}$ **(c)** $\frac{1}{4}$ **(d)** 12   **(e)** $\frac{1}{9}$ **(f)** $\frac{2}{3}$

2 Work out:

    (a) $4 \div \frac{1}{3}$ **(b)** $5 \div \frac{1}{4}$ **(c)** $6 \div \frac{2}{3}$ **(d)** $8 \div \frac{4}{5}$ **(e)** $9 \div \frac{3}{4}$

    (f) $\frac{1}{2} \div \frac{1}{4}$ **(g)** $\frac{2}{3} \div \frac{1}{3}$ **(h)** $\frac{3}{4} \div \frac{1}{4}$ **(i)** $\frac{1}{4} \div \frac{1}{8}$ **(j)** $\frac{2}{5} \div \frac{1}{10}$

3 Jack is preparing for a party. He estimates that each person will eat $\frac{1}{4}$ of a pizza.
How many people can he feed with 5 pizzas?

4 Fiona's favourite television programme lasts for $\frac{3}{4}$ of an hour.
She records each episode.

    (a) How many episodes can she record on a 3 hour tape?

    (b) How many can she record on a 2 hour tape? (Your answer will be a mixed number.)

5 Richard walks $\frac{2}{5}$ of a kilometre to school each day.
It takes him 10 minutes.

    (a) Write 10 minutes as a fraction of an hour.

    (b) Write down a division involving fractions that can be used to work out Richard's speed in kilometres per hour.

    (c) Work out Richard's speed giving your answer as a mixed number.

6 Work out:

> *Remember BIDMAS.*
> *Work out inside the brackets first.*

    (a) $\left(\frac{2}{3} \times \frac{1}{2}\right) \div 4$ **(b)** $\left(\frac{2}{7} \times 14\right) \div \frac{1}{2}$ **(c)** $\left(\frac{1}{3} \times \frac{1}{4}\right) \div \frac{1}{2}$ **(d)** $\left(\frac{4}{9} \div \frac{2}{3}\right) \times 3$

    (e) $\left(\frac{4}{7} \div \frac{3}{7}\right) \times \frac{3}{4}$ **(f)** $\left(\frac{5}{8} \div \frac{5}{16}\right) \times \frac{1}{4}$ **(g)** $\left(\frac{3}{10} \div \frac{3}{20}\right) \div 2$ **(h)** $\left(\frac{7}{12} \div \frac{1}{6}\right) \div \frac{1}{2}$

Investigation   Work out

    (a) $\frac{1}{2} \times 2$ **(b)** $5 \times \frac{1}{5}$ **(c)** $\frac{3}{4} \times \frac{4}{3}$

    (d) $\frac{2}{7} \times \frac{7}{2}$ **(e)** $\frac{2}{3} \times \frac{3}{2}$

    What do you notice about your answers?
    What is the effect of multiplying a number by its reciprocal?
    Is this always true?
    Explain your answer.

Finishing off

Now that you have finished this chapter you should be able to:

- multiply fractions
- find a fraction of an amount
- divide a fraction by a whole number
- find the reciprocal of a fraction or a whole number
- divide by a fraction.

Review exercise

1 Copy and complete this multiplication table.

×	$\frac{1}{2}$	$\frac{3}{7}$	$\frac{7}{9}$		$\frac{1}{5}$		$\frac{1}{3}$
$\frac{1}{2}$							
$\frac{3}{4}$							
$\frac{2}{5}$							
$\frac{5}{8}$							
			$\frac{2}{3}$				
$\frac{5}{16}$			$\frac{3}{16}$				
					$\frac{1}{15}$	$\frac{1}{4}$	$\frac{1}{9}$

2 Work out the following.
You should be able to do them in your head.

(a) $\frac{3}{4}$ of 20   (b) $\frac{2}{5}$ of 20   (c) $\frac{5}{9}$ of 27   (d) $\frac{2}{15}$ of 30

(e) $\frac{4}{7}$ of 28   (f) $\frac{3}{10}$ of €2.50   (g) $\frac{2}{3}$ of €3.99   (h) $\frac{4}{5}$ of €3.60

3 Calculate the area of the following rectangles

(a)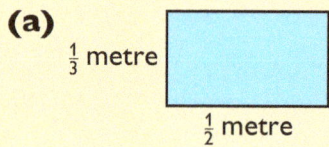
$\frac{1}{3}$ metre

$\frac{1}{2}$ metre

(b)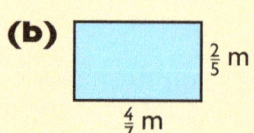
$\frac{2}{5}$ m

$\frac{4}{7}$ m

(c)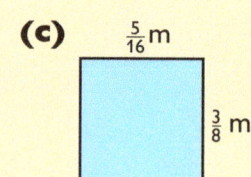
$\frac{5}{16}$ m

$\frac{3}{8}$ m

4 Calculate the areas of the following triangles

(a)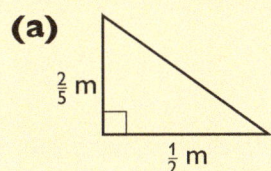

(b)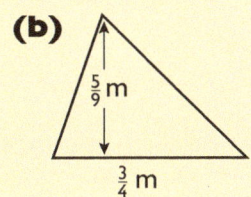

(c)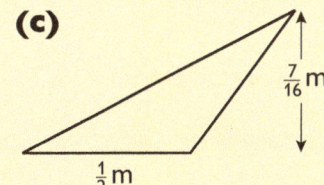

5 The area of each of the shapes below is given.
For each shape find the missing length.

(a)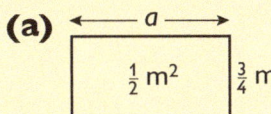

(b)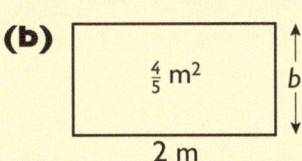

(c)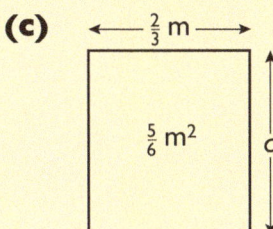

(d)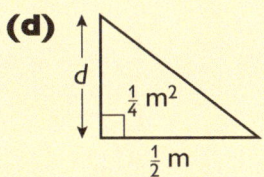

(e)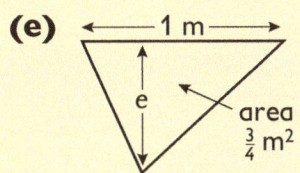

(f)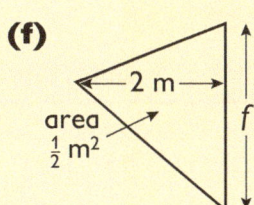

6 Jeremy has drawn this
pie chart to show how he
spends his pocket money.

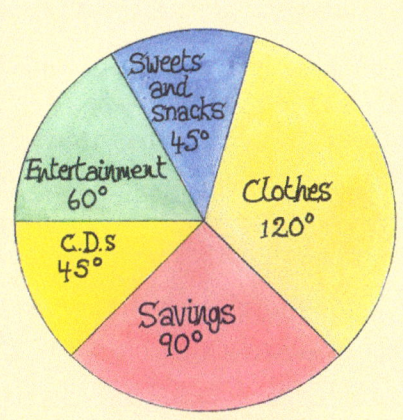

(a) What fraction of his money
does he spend on

  (i)   entertainment

  (ii) sweets and snacks

  (iii) clothes?

(b) He gets a total of €600 pocket-money
each year. (€50 each month.)
Work out how much he spends on
each category in a year.

7 Work out the following speeds

(a) $\frac{4}{5}$ of a kilometre in $\frac{3}{4}$ of an hour

(b) $\frac{3}{10}$ of a kilometre in $\frac{4}{5}$ of an hour

(c) $\frac{5}{8}$ of a kilometre in $\frac{1}{3}$ of an hour

(d) 6 kilometres in $\frac{1}{10}$ of an hour.

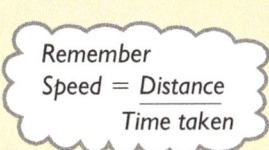

Remember
$Speed = \dfrac{Distance}{Time\ taken}$

Do you remember?

A **transformation** maps an object (a shape) onto its image.
Translations, reflections, rotations and enlargements are all transformations.

Translation

Object PQR is translated to the **image** P′Q′R′.
The object slides *without* turning.

- What can you say about the paths of the points?
- What can you say about the size and shape of the object and the image?
- How do you give instructions for a translation?
- What are the instructions for *this* translation?

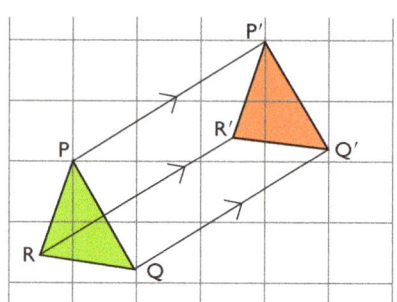

Reflection

Object DEF is reflected in the line XY. The image is D′F′E′.

- What can you say about the distances of the object and the image from the mirror line?
- What can you say about the size and shape of the object and the image?
- Why has the order of the letters changed?

In a reflection the object and image are
the same shape and size.
We say they are **congruent**.

For 'same size and shape' you might have said identical. From now on we say congruent.

Task

The triangles on this grid can be **mapped** on to each other.

Examples

C to E Translation 4 to the right and 3 down.
F to E Reflection in the x axis.

Describe *fully* the following transformations.

1 A to E	**2** B to C	**3** D to B
4 G to C	**5** A to B	**6** A to D
7 B to A	**8** D to C	**9** H to A
10 E to A	**11** D to A	**12** A to H

- Find two transformations which, one after the other, take B to G.
- Does the order of these two transformations matter?

- Which points do not move in a reflection?

Exercise

1 Look at this wallpaper pattern of repeated crescents.

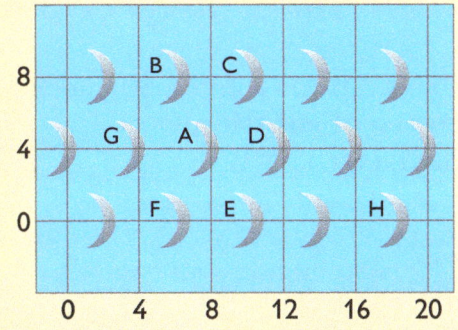

Crescent A is mapped to crescent C by a translation of 2 to the right and 4 up.
Describe fully the translations which take

(a) A to B **(b)** A to E **(c)** F to A

(d) E to B **(e)** E to C **(f)** D to G

(g) B to H **(h)** H to G

2 Draw and label x and y axes from -10 to $+10$.

(a) Draw the hexagon PQRSTU where P is $(3,10)$, Q $(5,10)$, R $(6,7)$, S $(6,5)$, T $(3,5)$ and U $(2,7)$.

After a translation, P is mapped to P′ $(-3,4)$.

(b) Write down the instructions for this translation.

(c) Calculate and write down the co-ordinates of Q′, R′, S′, T′ and U′.

(d) Plot and draw image P′Q′R′S′T′U′.

P′Q′R′S′T′U′ is translated 2 to the right and 7 down.
The image is P″Q″R″S″T″U″.

(e) Write down the co-ordinates of P″, Q″, R″, S″, T″, U″.

(f) Calculate the single transformation PQRSTU → P″Q″R″S″T″U″.

(g) Plot and draw P″Q″R″S″T″U″.

3 Draw and label x and y axes from -7 to $+7$.

(a) Plot and label the following points.
Join the points to form quadrilateral WXYZ.

> W $(2,1)$ X $(3,5)$ Y $(6,6)$ Z $(7,2)$

(b) **(i)** Reflect WXYZ in the x axis to form image W′X′Y′Z′.

   (ii) Write down the co-ordinates of W′, X′, Y′, and Z′.

   (iii) Describe the change to the co-ordinates of an object under a reflection in the x axis.

(c) **(i)** Reflect W′X′Y′Z′ in the y axis to form image W″X″Y″Z″.

   (ii) Write down the co-ordinates of W″, X″, Y″ and Z″.

   (iii) Describe the change to the co-ordinates of an object under a reflection in the y axis.

Activity

 Why do emergency vehicles have writing like this on the front?

Write a message which can be read when reflected in a mirror.

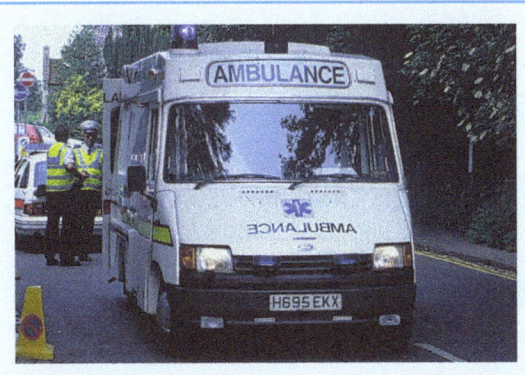

Rotation

Object ABC is rotated 60° clockwise about the
centre of rotation O.
The image is A′B′C′.

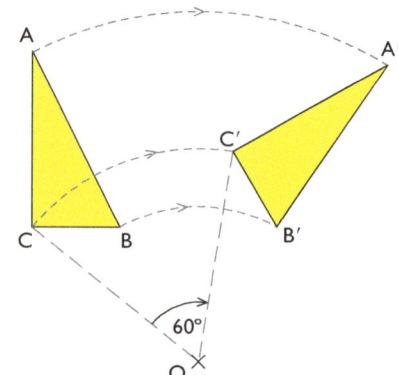

 **What can you say about the distances of
the object and the image from the centre of rotation?**

The path of each point is an **arc**.

 What fraction of a circle is each arc in this rotation?

Task

These shapes can be rotated onto each other.

Example

> J to K Anticlockwise rotation of 270°
> about the origin.

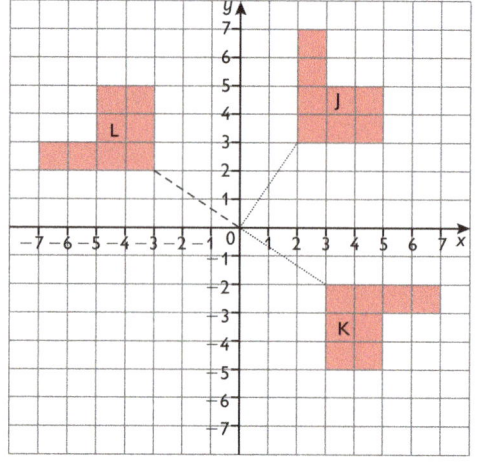

1 **(a)** Describe fully the following rotations.
   (i) J to L   **(ii)** L to J   **(iii)** K to L
   (iv) L to K   **(v)** K to J.

   (b) Compare your answers with a friend.
   Are they the same?
   How can they be different?

2 Ask your teacher for a copy of the grid or copy it onto squared paper.
   (a) Reflect Shape L in the *x* axis and label the image R.
   (b) Reflect R in the *y* axis. Where does it land?
   (c) Which single transformation is **equivalent** to a reflection in the *x* axis
   followed by a reflection in the *y* axis?

 **What is the meaning of the word congruent?
Look at the four shapes in the Task.
Are they congruent?**

 Can any of the four shapes be translated onto each other?

 How can two different rotations be equivalent to each other?

 **How can you map an object onto itself?
Think of two different transformations which take J to J.**

Exercise

1 **(a)** Draw and label x and y axes from -10 to $+10$.
Plot $(2, 4)$, $(4, 6)$ and $(2, 9)$ and join them to form a triangle.
Label the triangle A.

(b) Rotate A clockwise through 90° about the origin. Label the image B.

(c) Rotate B anticlockwise through 180° about the origin. Label the image C.

(d) Write down *two* transformations which will map C onto A.

2 Look at the picture of a spanner turning a bolt.
The bolt head is a regular hexagon.
Point C is the centre of rotation.

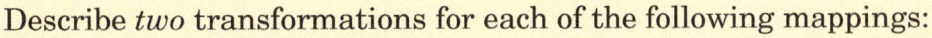

(a) The spanner is rotated 60° clockwise about C.
What can you say about the *appearance* of the bolt before and after the rotation?

(b) Write down three other clockwise rotations which do not change the appearance of the bolt.

(c) Write down two anticlockwise rotations which do not change the appearance of the bolt.

(d) Give a *full* description of the symmetry of the bolt.

3 For this question you will need a protractor and a pair of compasses.
Here is a plan of a children's roundabout.
O and T show the starting positions of Owain and Tom.
Make a copy of the diagram.
On your diagram show the positions of Owain
and Tom after each of the following
transformations from the starting positions:

(a) Clockwise rotation of 90° about C.
Label them O_1 and T_1.

(b) Clockwise rotation of 120° about C.
Label them O_2 and T_2.

(c) Clockwise rotation of 290° about C.
Label them O_3 and T_3.

Describe *two* transformations for each of the following mappings:

(d) O_1 to O_2 **(e)** T_1 to T_3 **(f)** O_2 to O   **(g)** O_3 to O_1

Investigation

Draw and label x and y axes from -8 to $+8$.
Plot and label the following points and join them to form quadrilateral ABCD.

A (2,1)   B (4,0)   C (7,5)   D (4,−2)

1 Rotate ABCD clockwise through 90° about the origin. Label the image A′B′C′D′.
2 Rotate ABCD through 180° about the origin. Label the image A″B″C″D″.

Compare the object co-ordinates and corresponding image co-ordinates.
Explain how the rotations about the origin change the co-ordinates.

Enlargement

In this diagram object ABCDE has been **enlarged** to image A′B′C′D′E′.

In what way are the object and image different?

In what way are the object and image the same?

Are the object and image congruent?

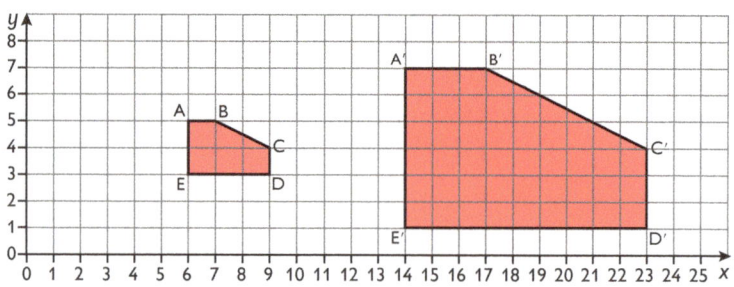

Task

Ask your teacher for a copy of the diagram or copy it onto centimetre squared paper.

1 Draw a straight line **(a)** through A and A′   **(b)** through E and E′.

Extend these lines so that they intersect. Label the intersection P.

P should be at the point (2, 4).

Draw the lines through the other **corresponding** object and image points.
P is the **centre of enlargement**.

2 Copy and complete the table below.
Measure the distances in centimetres.

Distance from centre to object point	Distance from centre to corresponding image point	Image distance ÷ object distance (to one decimal place)
PA =	PA′ =	PA′ ÷ PA =
PB = 5.1 cm	PB′ = 15.3 cm	PB′ ÷ PB =
PC =		
PD =		
PE =		

3 Copy and complete the table below.

Object length	Corresponding image length	Image length ÷ object length
AE = 2.0 cm	A′E′ = 6.0 cm	A′E′ ÷ AE =
ED =	E′D′ =	
AD =		
CD =		

The image is 3 times further *from the centre of enlargement* than the object.
All image lengths are 3 times the corresponding object lengths.
This enlargement is **scale factor** 3.

Corresponding angles in the object and image are equal.
Check with a protractor.
The object and image are **similar shapes**.

Exercise

1 Draw and label x and y axes from 0 to 15.
Label the origin O.
Plot and label the points A(3, 2), B(3, 7), C(4, 7), D(4, 3), E(7, 3) and F(7, 2).
Join them to form an L shape.

(a) With the origin as centre, enlarge object ABCDEF by a scale factor of 2.
Label the image A′B′C′D′E′F′.

(b) Write down the co-ordinates of A′, B′, C′, D′, E′ and F′.
Compare them with the object co-ordinates. What do you notice?

2 The diagram below shows a film being projected.
An object on the film is enlarged to form its image on the screen.
The light source is the centre of enlargement.

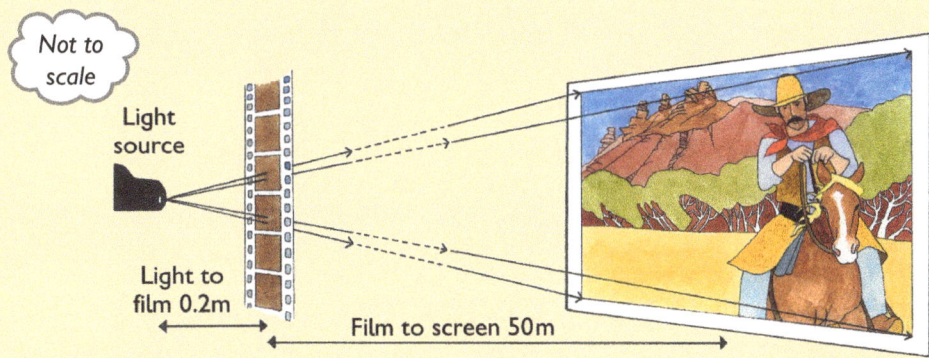

(a) Look at the distances on the diagram.
Explain why the scale factor of the enlargement is 251.

(b) The width of the picture on the film is 16 mm.
What is the width of the picture on the screen?
Give your answer in metres.

(c) On the screen, the cowboy's hat is 1 m wide.
How wide is the hat on the film? Give your answer to the nearest mm.

3 The co-ordinates of a polygon are P (2,5), Q (3,9), R (8,7) and S (9,2).
After an enlargement with the origin as centre, P is mapped to P′ (8,20).

(a) What is the scale factor?

(b) Write down the co-ordinates of Q′, R′, S′.

With the origin as centre, PQRS is enlarged to form image P″Q″R″S″.
Write down the co-ordinates of P″, Q″, R″ and S″ for each of the following
scale factors.

(c) 5   **(d)** 10   **(e)** 20   **(f)** 100

4 Draw and label a pair of axes, with x from -7 to 10 and y from -4 to 10.
Draw a rectangle with vertices at $(-2,3)$, $(3,3)$, $(3,-1)$, $(-2,-1)$.
Using the origin as centre, enlarge the rectangle with a scale factor of

(a) 2   **(b)** 3

Finishing off

Now that you have finished this chapter you should know:

- and recognise full descriptions of translations, reflections, rotations and enlargements
- and understand the meanings of congruent and similar
- that in a translation, rotation and reflection the object and image are congruent
- that in an enlargement the object and image are similar.

Review exercise

Use this grid for Questions 1 and 2.

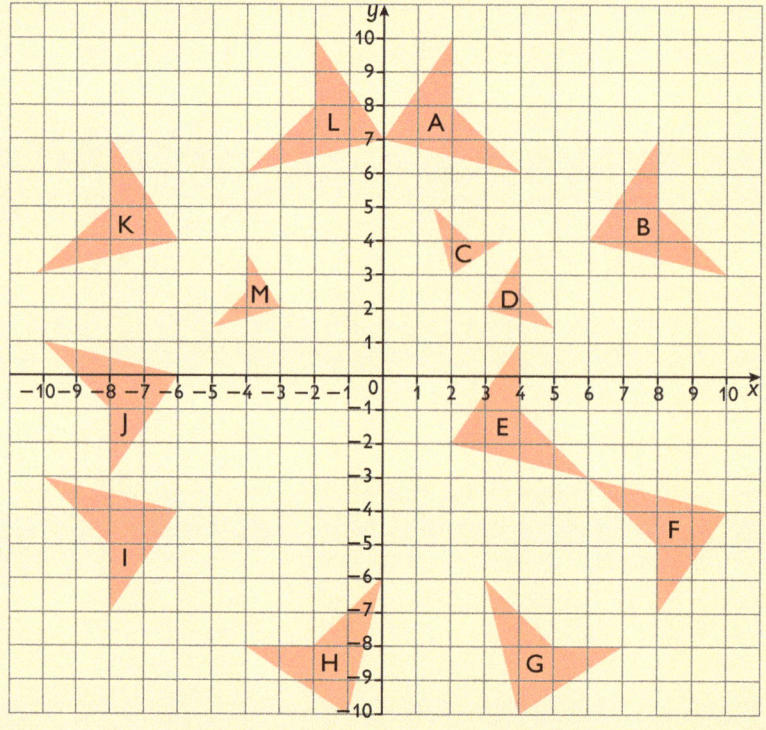

1. Look at the shapes on the co-ordinate grid above.
   Describe fully the transformations for the following mappings.
   (a) K to L   **(b)** A to B   **(c)** K to B   **(d)** I to B
   (e) D to B   **(f)** K to I   **(g)** C to M   **(h)** F to E
   (i) F to I   **(j)** A to L   **(k)** M to K   **(l)** H to G
   (m) C to D   **(n)** J to I   **(o)** F to J

2. Look at the shapes on the grid. State a mapping
   for each of the following transformations.
   (a) Reflection in the line $y = -x$
   (b) Reflection in the line $y = 2$
   (c) Rotation of 180° about (8, 0)

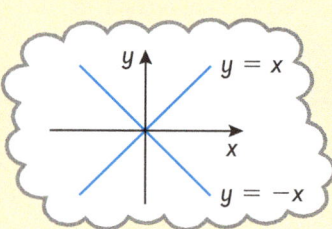

3 Look at this diagram.

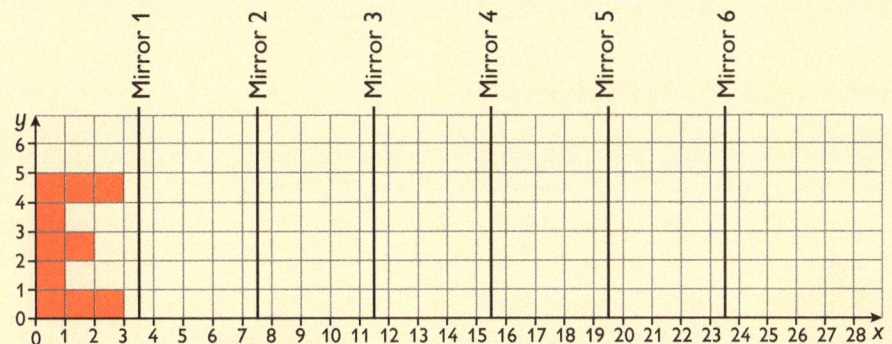

(a) The E shape has 12 vertices. Write down the co-ordinates of the vertices.
(b) The equation of Mirror Line 1 is $x = 3\frac{1}{2}$.
    What are the equations of the other 5 mirror lines?
(c) Copy the diagram onto squared paper.
(d) Reflect the E shape in mirror 1.
(e) Reflect the image from part (d) in mirror 2.
(f) Reflect the image from part (e) in mirror 3.
(g) Continue the process with mirrors 4, 5 and 6.
(h) Describe what is happening.

4 A shape is translated five times:
8 to the right and 4 down *followed by* 3 to the left and 22 down *followed by*
43 down *followed by* 17 to the right and 16 up *followed by* 3 down.
Calculate the translation which takes the shape back to its starting place.

5 (a) Copy the diagram onto a
    sheet of paper as shown.
(b) Use a ruler to construct an
    enlargement of XYZ, scale
    factor 2, centre C.
    Label the image X′Y′Z′.
(c) Construct an enlargement
    of XYZ, scale factor 3, centre C.
    Label the image X″Y″Z″.
(d) What transformation
    maps X′Y′Z′ onto X″Y″Z″?

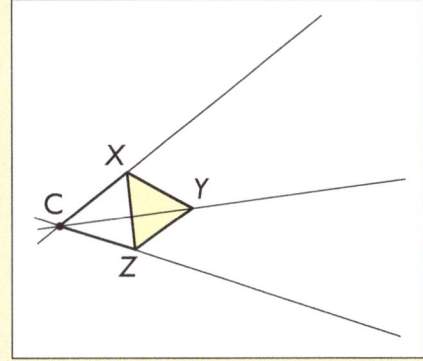

6 After a rotation is repeated 8 times the object has returned to its starting
position.
What is the angle of the rotation?
Is there more than one possible answer?

Investigation
 1 Write down two different rotations where the object ends in its
   starting position.
 2 What can you say about an enlargement where the object is
   unchanged?
 3 Can a reflection leave an object unchanged?
 4 Write down a translation where the object does not move.

25 Getting the most from your calculator

Multi-stage calculations

Daniel and Jo use their calculators to work out

$$\frac{2.2}{4.4 + 1.1}$$

No it's 1.6

The answer is 0.4

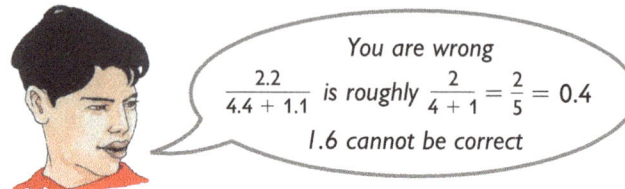

You are wrong
$\frac{2.2}{4.4 + 1.1}$ is roughly $\frac{2}{4 + 1} = \frac{2}{5} = 0.4$
1.6 cannot be correct

 How was the wrong answer obtained?

Many calculations require part of the answer to be found before the rest can be worked out.
This can be done on your calculator by using brackets, using the memory or using the answer button.

Only newer calculators have this.

" Do the right thing!

Use your calculator to work out $\frac{3.9}{4.3 - 1.6}$. Here are three different methods.

1 Using brackets `3.9` `÷` `(` `4.3` `−` `1.6` `)` `=`

2 Using the memory work out $4.3 - 1.6$ first `4.3` `−` `1.6` `=`

Store this in the memory by pressing `M+` or `STO`

Clear the screen `AC`

Now work out `3.9` `÷` `MR` or `RCL` `=`

3 Using the answer button (not all calculators have this)
Work out $4.3 - 1.6$ `4.3` `−` `1.6` `=`

⚠ Do *not* clear the screen.

Now work out `3.9` `÷` `ANS` `=`

 Explain why the calculator sequence `2.4` `+` `3.1` `÷` `4.2` `×` `1.6` `=`

will not give the correct answer to $\frac{2.4 + 3.1}{4.2 \times 1.6}$.

 **Use two of the methods above to obtain the correct answer.
Check your answer with an approximation.**

Exercise

1 Using the brackets on your calculator work out the following:

(a) $2.1 \times (3.9 + 8.7)$

(b) $3.4 \div (8.2 + 4.6)$

(c) $(67 + 93) \times (104 - 67)$

(d) $(8.1 + 3.6)^2$

(e) $(5.9 + 3.2) - (1.1 + 2.7)$

(f) $\dfrac{86 + 72}{16}$

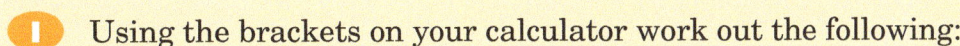

Remember **BIDMAS**
Brackets
Index
Divide
Multiply
Add
Subtract

2 For each of the following
(i) find an approximate answer
(ii) use the memory or answer button to calculate the exact answer.

(a) $\dfrac{64}{2.7 + 1.3}$

(b) $\dfrac{83.6 + 42.4}{1.4 \times 1.5}$

(c) $\dfrac{82 \times 7}{6.1 - 4.5}$

(d) $\dfrac{102 + 364}{434 - 178}$

(e) $\frac{1}{2}$ of $(62 + 97)$

(f) $\frac{3}{4}$ of $(8.2 + 4.5 + 7.2)$

3 Each of the following calculations is wrong.
In each case **(i)** find the correct answer
(ii) explain how the wrong answer was obtained.

(a) $\dfrac{45 + 36}{1.8 + 3.6} = 48.6$

(b) $\dfrac{45 + 36}{1.8 + 3.6} = 68.6$

(c) $(1.2 + 3.6)^2 = 14.16$

(d) $\dfrac{14 - 6.8}{25} = 13.728$

Activity

$16 \times 25 = 400$
I can work it out more quickly in my head than on a calculator. Just add 2 noughts and divide by 4.

Ali

Working with a friend, one of you use a calculator and the other use Ali's method.

Work out 8×25 12×25 6×25 25×25
Who is quicker?
1 Suggest a quick way to multiply by 5, 50 and 250.
2 Write a multiplication sum involving these numbers.
3 Challenge a friend to work more quickly with a calculator.

Investigation

Do the sum $7 + 3 \div 3 - 1$ on your calculator.
Explain the order for the operations.

The answer is 7.

Work out   **(a)** $7 + 3 \div (3 - 1)$ **(b)** $(7 + 3) \div (3 - 1)$
Place the brackets in different places.
How many different answers can you get?

Do not move the positions of the numbers or the operations $+$, $\div$ and $-$.

What happens when you place brackets in the following:
1 $9 - 6 \div 3 + 1$ **2** $8 + 4 \times 2 - 1$

Fractions

Find the fraction button on your calculator.

Press 3 $a\frac{b}{c}$ 4

Your calculator shows $\frac{3}{4}$.

> A mixed number contains a whole number and a fraction.

Press 2 $a\frac{b}{c}$ 4 $a\frac{b}{c}$ 5

Your calculator shows the mixed number $2\frac{4}{5}$.

Enter these fractions into your calculator.

$$\frac{1}{2} \qquad \frac{2}{5} \qquad \frac{5}{12} \qquad \frac{11}{25} \qquad 1\frac{2}{3} \qquad 7\frac{4}{13}$$

 Write down the display each time.

> Clear the screen between each fraction.

Task

1 Work out the following without using your calculator.

$$\frac{1}{2} + \frac{1}{3} \qquad \frac{5}{7} - \frac{3}{14} \qquad \frac{3}{5} \times \frac{2}{9} \qquad 5 \div \frac{2}{3}$$

2 Now check you get the same answers on your calculator.

Task

1 Enter $\frac{4}{5}$ into your calculator. Press = $a\frac{b}{c}$. What happens?

Press $a\frac{b}{c}$ again. What happens this time?

2 Use your calculator to convert these fractions to decimals.

$$\frac{3}{10} \qquad \frac{9}{16} \qquad \frac{3}{7} \qquad \frac{2}{9}$$

? **When is your calculator able to convert a decimal to a fraction?**

Task

For the following

(a) Put the fraction into your calculator

(b) Press the = button   **(c)** Record the answer.

> Clear the calculator between each fraction.

1 $\frac{3}{6} =$ 2 $\frac{14}{49} =$ 3 $\frac{32}{48} =$

The calculator cancels the fractions down.

 For each cancelled down fraction above, write down the common factor for the top and bottom number.

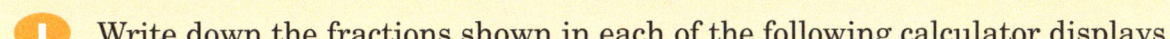

Exercise

1 Write down the fractions shown in each of the following calculator displays.

(a) `2 ⌐ 5`   (b) `11 ⌐ 12`   (c) `1 ⌐ 1 ⌐ 2`

(d) `2 ⌐ 7 ⌐ 9`   (e) `10 ⌐ 1 ⌐ 5`   (f) `4 ⌐ 9 ⌐ 11`

2 Cancel down the following fractions as far as possible.

(a) $\frac{5}{10}$ (b) $\frac{3}{9}$ (c) $\frac{10}{15}$ (d) $\frac{36}{90}$

Use your calculator to check your answers.

3 Using a calculator work out the following.

(a) $\frac{3}{5} + \frac{7}{9}$ (b) $\frac{8}{15} - \frac{2}{7}$ (c) $\frac{4}{7} \div \frac{5}{14}$ (d) $\frac{3}{4} \times \frac{7}{8}$

(e) $\frac{2}{3} \times (\frac{1}{4} + \frac{1}{2})$ (f) $(\frac{2}{5} + \frac{7}{9}) \times (\frac{1}{4} - \frac{1}{5})$ (g) $(\frac{3}{8} - \frac{1}{5}) \div \frac{1}{7}$

Investigation Enter the mixed number $1\frac{2}{3}$ into your calculator.

Press `SHIFT` or `2nd`   `a b/c`   Write down the display.

Why do you think the term top-heavy is used?

This is the top-heavy fraction $\frac{5}{3}$. It is equivalent to $1\frac{2}{3}$.

Use your calculator to convert the following to top-heavy fractions.

1 $1\frac{4}{5}$ **2** $2\frac{1}{3}$ **3** $3\frac{4}{7}$ **4** $4\frac{1}{4}$

Look carefully at your answers.
Write a set of instructions to convert a mixed number to a top-heavy fraction.
Ask a friend to test your instructions on the mixed number $5\frac{3}{4}$.

Investigation Work out $7 \div 2$ and $7 \times \frac{1}{2}$.

 You should get the same answer. Why?

Write down a multiplication that is equivalent to each of the following:

1 $5 \div 4$ **2** $11 \div 5$ **3** $15 \div 7$ **4** $22 \div 6$

Work out   **5** $9 \div 4 \times 3$ **6** $9 \times 3 \div 4$ **7** $\frac{3}{4} \times 9$

Why do 5, 6 and 7 all give the same answer?

For each of the following write down two equivalent calculations.

8 $16 \div 5 \times 4$ **9** $\frac{4}{7} \times 12$ **10** $22 \times 5 \div 11$

Check all your answers on a calculator.

Indices

To work out 2^7

Press $\boxed{2}$ then $\boxed{y^x}$ then $\boxed{7}$ and $\boxed{=}$

$\boxed{x^y}$ (above)

$\boxed{\wedge}$

This means
$2 \times 2 \times 2 \times 2 \times 2 \times 2 \times 2$
I will lose count of the number of 2s
if I use my calculator.
There must be an easier way.

Your calculator should have
one of these buttons. It is the
'power of' or index button.

? **What is the value of 2^7?**

Explain why this is the 7th term of the following sequence.   *2, 4, 8, 16, ..., ...*

Continue the sequence to the 7th term to check you get the same answer.

Task

These are cube numbers: 1, 8, 27, 64, ...
Write each number as $?^3$.
For example $8 = 2^3$.

These are powers of 4: 4, 16, 64, 256, ...
Write each number as $4^?$.
For example $16 = 4^2$.

For each sequence above
1 Find the next number in the sequence
2 Write an expression, using index notation, for the 10th number
3 Use your calculator to find the 10th number.
Explain why it is easier to use the index button on your calculator to find the 10th number.

Finding roots

Press $\boxed{\text{SHIFT}}$ or $\boxed{\text{2nd}}$ then the index button.

This works out the reverse or *inverse* of an operation.
It finds the root of a number, for example a cube or fourth root.

? **Write down the meaning of the fifth root of a number, in simple English.**

Press the following sequence into your calculator:

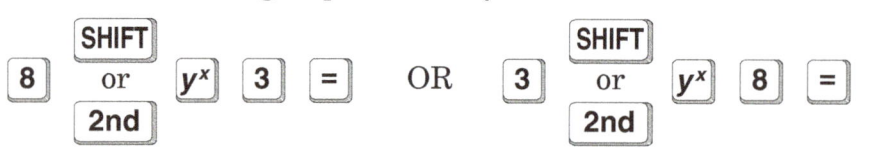

$\boxed{8}$ $\boxed{\text{SHIFT}}$ or $\boxed{\text{2nd}}$ $\boxed{y^x}$ $\boxed{3}$ $\boxed{=}$ OR $\boxed{3}$ $\boxed{\text{SHIFT}}$ or $\boxed{\text{2nd}}$ $\boxed{y^x}$ $\boxed{8}$ $\boxed{=}$

Find which of these
is the correct sequence of
keys for your calculator.

You have found $\sqrt[3]{8}$. This is the *cube root* of 8.

? **Using your calculator show that the fourth root of 81 is 3.**
What operation is performed by $\boxed{\text{SHIFT}}$ $\boxed{x^2}$ $\boxed{=}$ **?**

$\sqrt[4]{81} = 3$

Exercise

1 Work out
   (a) 6^4 **(b)** 30^4 **(c)** 3.6^5 **(d)** 1.4^6 **(e)** $\sqrt[3]{125}$ **(f)** $\sqrt[4]{1296}$

2 Write the following using index notation then evaluate using your calculator.
   (a) $5 \times 5 \times 5 \times 5 \times 5$ **(b)** $7 \times 7 \times 7 \times 7$ **(c)** $35 \times 35 \times 35$
   (d) $1.9 \times 1.9 \times 1.9 \times 1.9 \times 1.9$ **(e)** $2.6 \times 2.6 \times 2.6 \times 2.6 \times 2.6 \times 2.6$

3 Use your calculator to find the values of
   (a) 20^3 **(b)** 20^4 **(c)** 20^5 **(d)** 20^6
   Compare these answers with the answers to 2^3, 2^4, 2^5 and 2^6.

4 Work out the exact value of
   (a) 20^{10} **(b)** 20^{15}

   Hint: work out 2^{10} and 2^{15} first.

5 David is planning a six-week training schedule.

Activity	Week 1	Week 2	Week 3	Week 4	Week 5	Week 6
Sit-ups	20		20×1.2^2			
Swimming	10 lengths		10×1.2^2			
Jogging	5 km					
Exercise bike	15 minutes					

He has entered the first week of his programme in the table.
He aims to increase each activity every week, by multiplying his targets for the previous week by 1.2.
   (a) Work out his targets for week 2.
   (b) Explain how the calculations 20×1.2^2 and 10×1.2^2 give his targets in week 3 for sit-ups and swimming.
   (c) Write down the calculations that give his targets in week 3 for jogging and the exercise bike.
   (d) Work out all his targets for week 3. Give your answers to the nearest whole number.
   (e) **(i)** Write down similar calculations for all activities in week 6.
        (ii) Work out these values.
        Give your answers to the nearest whole number.

6 The numbers opposite are square numbers or cube numbers.

   216 64 225 4913 4096 289

   (a) Which two numbers are both a square number and a cube number?
   (b) Find another number that is both square and cube.

Activity   Find the value of the terms in the sequence
   Which is the first term that cannot be displayed as an ordinary number on your calculator?
   Write down the value of 0.1^{15}.

   $0.1, 0.1^2, 0.1^3, 0.1^4$

Finishing off

Now that you have finished this chapter you should be able to:

- use brackets on your calculator
- use the memory or an answer button
- use the fraction button
- use 'the power of' or index button.

Review exercise

1 Work out the following on your calculator.

(a) $(4.3 + 3.7)^2$ **(b)** $\dfrac{18}{(3.7 - 2.8)^2}$ **(c)** 87^4

(d) 7×1.4^5 **(e)** $\sqrt{(0.42 + 1.02)}$ **(f)** $\sqrt[3]{0.027}$

2 For each of the calculator sequences below
(i) write down the calculation that will be performed
(ii) do the calculation

(a) $\boxed{42}\ \boxed{\times}\ \boxed{(}\ \boxed{87}\ \boxed{-}\ \boxed{63}\ \boxed{)}\ \boxed{=}$

(b) $\boxed{65}\ \boxed{+}\ \boxed{18}\ \boxed{=}\ \boxed{M+}\ \boxed{AC}\ \boxed{14}\ \boxed{\times}\ \boxed{MR}\ \boxed{=}$

(c) $\boxed{(}\ \boxed{8.7}\ \boxed{+}\ \boxed{3.9}\ \boxed{)}\ \boxed{x^y}\ \boxed{4}\ \boxed{=}$

(d) $\boxed{3}\ \boxed{a\frac{b}{c}}\ \boxed{4}\ \boxed{+}\ \boxed{2}\ \boxed{a\frac{b}{c}}\ \boxed{2}\ \boxed{a\frac{b}{c}}\ \boxed{3}\ \boxed{=}$

(e) $\boxed{1.9}\ \boxed{+}\ \boxed{0.6}\ \boxed{=}\ \boxed{M+}\ \boxed{AC}\ \boxed{(}\ \boxed{2.9}\ \boxed{-}\ \boxed{1.5}\ \boxed{)}\ \boxed{\div}\ \boxed{MR}\ \boxed{=}$

(f) $\boxed{1.9}\ \boxed{+}\ \boxed{0.6}\ \boxed{=}\ \boxed{M+}\ \boxed{AC}\ \boxed{2.7}\ \boxed{-}\ \boxed{1.3}\ \boxed{\div}\ \boxed{MR}\ \boxed{=}$

(g) $\boxed{5}\ \boxed{x^y}\ \boxed{7}\ \boxed{\div}\ \boxed{5}\ \boxed{x^y}\ \boxed{6}\ \boxed{=}$

(h) $\boxed{2}\ \boxed{a\frac{b}{c}}\ \boxed{3}\ \boxed{x^y}\ \boxed{4}\ \boxed{=}$

3 For parts **(c)** and **(f)** of Question 1:
(i) Write calculator sequences similar to Question 2.
(ii) Check the sequences work on your calculator.

4 **(a)** Use your calculator to convert the following fractions to decimals.

(i) $\frac{3}{4}$ **(ii)** $\frac{1}{3}$ **(iii)** $\frac{3}{8}$ **(iv)** $\frac{7}{100}$ **(v)** $\frac{1}{80}$

(b) Find fractions that are equivalent to:

(i) 0.1 **(ii)** 0.125 **(iii)** 0.04 **(iv)** 0.0375 **(v)** 0.005

Cancel your fractions if possible.

5 **(a)** Choose a number.

Work out (your number)^2 + 5

What sequence of operations converts your answer back to your original number?

(b) Repeat (a) for

(i) (your number − 8)^3 − 6 **(ii)** $\sqrt[3]{\text{(your number)}} \div 5$

6 **(a)** Complete the table.

Number	Square	Last digit
1	1	1
2	4	4
3		
4		
5	25	5
6		
7		
8		
9		

(b) Describe the pattern in the last column.
(c) How does it continue?
(d) Are there any numbers that do not appear in the last column?
(e) Explain why 167 cannot be a square number.

Do not try to find the square root.

Investigation

Look at this equation. It shows the powers of negative numbers.
Use the following sequence to work this out on your calculator:

$(-1)^2 = -1 \times -1 = 1$

[ ] [±] [1] [ )] [x^2] [=]

The ± button will enter a negative number.

Or [1] [±] [x^2]

1 Use your calculator to work out **(a)** $(-1)^3$ **(b)** $(-1)^4$ **(c)** $(-1)^5$
2 Write down **(a)** $(-1)^9$ **(b)** $(-1)^{100}$. Explain your answers.

Answers

1 Co-ordinates and graphs (pages 10–11)

1 A(1.6), 1.4) B(−2.4, −1.2) C(−0.8, −1.2) D(−0.8, −2.8)
2 (a) a(−1, 0) (0, 1) (1, 2) (2, 3)   (b) $y = x + 1$
   (a) b(−1, 3) (0, 2) (1, 1) (2, 0)   (b) $x + y = 2$
   (a) c(−2, 2) (−2, 1) (−2, 0) (−2, −1) (b) $x = −2$
   (a) d(1, −2) (2, −2) (3, −2)   (b) $y = −2$
3 (a)

x	−3	−2	−1	0	1	2	3
$3x$	−9	−6	−3	0	3	6	9
$−2$	−2	−2	−2	−2	−2	−2	−2
$y = 3x − 2$	−11	−8	−5	−2	1	4	7

(b)

x	−3	−2	−1	0	1	2	3
$4x$	−12	−8	−4	0	4	8	12
$+1$	+1	+1	+1	+1	+1	+1	+1
y	−11	−7	−3	1	5	9	13

(c) (i) Ask your teacher to check your diagram.
   (ii) −2, 1   (iii) (−3, −11)   (iv) 3, 4
4 (a) gradient = $\frac{4}{3}$, y-intercept = 2
   (b) gradient = 1, y-intercept = −1
5 $y = 5x − 3$ because the gradient, 5, is larger than the gradient of $y = 4x + 20$ which is 4.

Investigation
Ask your teacher to check your graphs.
(a) $y = x − 3$, $y = x + 4$ and $y = x + 2$.
   $y = 2x − 3$, $y = 4 + 2x$ and $y = 2x + 2$.
   The coefficient of x is the same size.
(b) $y = x − 3$, $y = 2x − 3$; $y = x + 4$, $y = 4 + 2x$; $y = x + 2$,
   $y = 2x + 2$
   The number terms in the equation are the same size.
(c) $y = 3x − 5$

2 Numbers (pages 20–21)

1 13.5 cm, 9.6 cm, 7.4 cm, 5.2 cm, 3.6 cm
2 10.77 s, 10.79 s, 10.81 s, 11.08 s
3 9.93 s, 9.95 s, 10.03 s, 10.2 s, 10.4 s
4 1.43×10^5, 1.2×10^5, 5.2×10^4, 4.8×10^4, 1.2756×10^4,
   1.21×10^4, 6.8×10^3, 4.9×10^3, 3.0×10^3
5 A 0.73,   B 0.78,   C 0.85
6 X 0.046,   Y 0.058,   Z 0.074
7 $25 < h < 35$

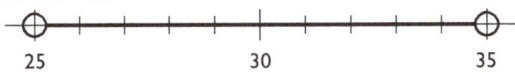

8 $60 < h < 75$

9 (a) 0.17 (b) 0.017 (c) 0.0017 (d) 0.17 (e) 0.017
   (f) 0.0017 (g) 0.17 (h) 0.017 (i) 0.0017
10 (a) 0.236 (b) 0.009 (c) 0.016 (d) 0.236 (e) 0.009
   (f) 0.016 (g) 0.236 (h) 0.009 (i) 0.016
11 The answer is smaller.
12 (a) 60 (b) 60 (c) 60 (d) 60 (e) 6 (f) 60
13 (a) 2000 (b) 4000 (c) 7300 (d) 2000 (e) 4000 (f) 7300
14 The answer is larger.
15 (a) (i) 1.8 h (ii) 0.9 h (iii) 3.2 h
   (b) (i) 1 hour 48 minutes (ii) 54 minutes
      (iii) 3 hours 12 minutes

3 Angles (pages 30–31)

1 A = 70° (Angles about a point add up to 360°)
   B = 49° (Vertically opposite angles are equal)
   C = 70° (Angles on a straight line add up to 180°)
   D = 73° (Angles in a triangle add up to 180°)
   E = 64° (Angles on a straight line add up to 180°)
   F = 65° (Alternate angles are equal)
   G = 39° (Corresponding angles are equal)
   H = 141° (Angles on a straight line add up to 180°)
   I = 45° (Angles about a point add up to 360°)
   J = 133° (Corresponding angles equal; adjacent angles add
         up to 180°)
   K = 47° (Corresponding angles equal; adjacent angles add
         up to 180°)
   L = 133° (Corresponding angles equal; adjacent angles add
         up to 180°)
   M = 72° (Exterior angle equals the sum of the opposite two
         interior angles; isosceles triangles; alternate
         angles are equal)
   N = 94° (Corresponding angles equal; alternate angles add
         up to 180°)
   P = 62° (Corresponding angles equal; alternate angles add
         up to 180°)
2 78°
3 (c) (i) 58° is acute (ii) 213° is reflex
      (iii) 90° is a right angle (iv) 111° is obtuse
      (v) 180° is a straight line (vi) 300° is reflex
      (vii) 354° is reflex
4 (a) (i) 031° (ii) 153° (iii) 326°
      (iv) 259° (v) 098° (vi) 270° (due West)
   (b) (i) 211° (ii) 333° (iii) 146°
      (iv) 079° (v) 278° (vi) 090° (due East)
   (c) Carham to Egwell and Filwood to Durton,
      Egwell to Carham and Durton to Filwood
5 A = 80°, B = 100°   6 X = 83°   7 261°
8 Split the pentagon into 3 triangles and $3 \times 180° = 540°$

4 Displaying data (pages 42–43)

1 (a)

Time spent playing computer games (min)	Tally	Frequency
0–50	IIII I	6
51–100	IIII IIII	9
101–150	IIII I	6
151–200	IIII	4
201–250	I	1
251–300	III	3
301–350	I	1

(b) Check diagram   (c) Students are not at school in
   August so some will spend more time playing computer
   games. The peak of the new bar chart will be further to
   the right in this case.
2 (a) 77, no (it is impossible)
   (b) Boys: median 5, mean 5   Girls: median 5, mean 5
      They rate them the same (c) Boys 5, girls 6
3 (a) Discrete (b) Discrete (c) Continuous
   (d) Continuous (e) Discrete
4 (a) 18 (b) 10 seconds
   (c) The table allows for notes which are sung for less than
      35 seconds (d) t stands for time (in seconds)
5 (d) Robert's
6 (b) Negative correlation (c) Clara. You cannot be sure.

5 Decimals (pages 52–53)

1 (a) 1.29 (b) 15.89 (c) 0.123 (d) 4.745 (e) 0.012
2 (a) 5.01 (b) 6.41 (c) 3.46 (d) 1.40 (e) 100
3 (a) 34.3 (b) 14.5 (c) 1.6 (d) 1.65
   (e) 0.14 (f) 1.0 (c) 0.50 (d) 1.00
4 (a) (i) $29 \times 2 = 58$ (ii) 57.816
   (b) (i) $30 \div 10 = 3$ (iii) 2.619827586
   (c) (i) $620 \times 0.04 = 24.8$ (iii) 24.7779
   (d) (i) $500 \times 30 \div 1000 = 15$ (iii) 16
5 (a) 7.020 litres (b) 16.58 hours (c) 12.267 minutes
   (d) 6.5833 days (e) €5.04
6 (a) €4 and 9 cent (b) 6 days 8.4 hours
   (c) 8 hours 18 minutes (d) 2 km 900 m
   (e) 3 minutes 24 seconds (d) 5 days 6 hours

7 (a) €19.81 (b) €3.03 (c) €4.47
8 (a) €1.66 (b) 90 g (c) 2 m (d) 4.8 km
9 (a) 3 (b) 4
10 (a) 3.8 cm (b) 15.2 cm. Rounding.

Investigation
1 0.7, 0.67, 0.667 0.4, 0.43, 0.429 0.5, 0.45, 0.455
0.3, 0.31, 0.308 0.2, 0.22, 0.222
2 Check findings

6 Using variables (pages 60–61)
1 (a) $7x + 12y$ (b) $5k + 9$ (c) $3a + 2b$ (d) -1
(e) $4s + 11t$ (f) $h - g$ (g) $7x$ (h) x (i) $20a$ (j) $4p + 7$
2 (a) $20f$ (b) s^2 (c) $5m^2$ (d) $3de$ (e) $4ab$ (f) $6g^2$
(g) $6x$ (h) $35y$ (i) p^3
3 (a) $10 + 5c$ (b) $3x - 18$ (c) $7p + 7r$ (d) $8d + 12$
(e) $4b + 32$ (f) $14r - 7$ (g) $3 - 3k$ (h) $10a - 15b$
(i) $5x + 5y$ (j) $6x - 3y$ (k) $18a - 24b$ (l) $36q - 60r$
4 (a) $6(3a + 1)$ (b) $4(3d - 1)$ (c) $12(2b - 3c)$
(d) $44(2h - 1)$ (e) $3(5k - 4)$ (f) $10(1 - 3f)$
(g) $12(a + 2b)$ (h) $12(a + 2b)$ (i) $4(x + 3y - 4z)$
5 (a) $p + 4p + 4p + 6p = 15p$ (b) (i) $360°$ (ii) $24°$
(c) $24°, 96°, 96°, 120°$
6 (a) $24n - 8$ (b) 64 cm (c) (i) 2 cm (ii) 2 cm
7 (a) $5c + 420$ where c is the cost in cent of a bottle of coke.
(b) 50 cent

Investigation
1

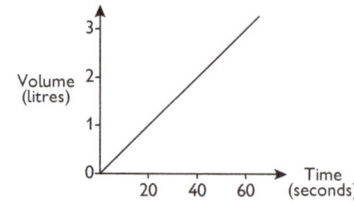

		Easy questions, e				
	0	**1**	**2**	**3**	**4**	**5**
0	0	600	1200	1800	2400	3000
1	1200	1800	2400	3000	3600	
2	2400	3000	3600	4200		
3	3600	4200	4800			
4	4800	5400				
5	6000					

Hard questions, h

Winnings Table (€)

2 $5 - (e + h)$ **3** Winnings $= 1200h + 600e$
4 Ask your teacher to check your answers.

7 Constructions (pages 72–73)
1 (a) Check diagrams. (iv) Not possible (b) (i) $31°, 52°, 97°$
(ii) $70°, 6.4$ cm, 7.8 cm (iii) $38°, 8.4$ cm, 5.3 cm
2 53 m **3** 9.9 cm **4** 6.9 cm, 10.6 cm
5 (b) $271°$ 109 km **6** They are perpendicular
7 (a) $149°$ (b) Yes (distance to hotel is 13.9 miles)

Investigation
The three perpendicular bisectors meet at a point.

Activity
1 and **2** Ask your teacher to check your diagram. **3** 96 m

8 Using graphs (pages 78–79)
1

(graph with Volume (litres) on vertical axis 0–3, Time (seconds) on horizontal axis 20, 40, 60; straight line from origin)

2 The peas are frozen which results in a drop in temperature initially. The temperature of the water increases but cannot exceed boiling point.
3 (a)–(c) Ask your teacher to check your graphs.
(d) The canoe and rowing boat lines are parallel with the canoe more expensive. The motor boat is the most expensive.
4 (a) 16 kg (b) $8\frac{1}{2}$ stones (c) Approx. 6 kg
5 (a) Paula stops for 15 minutes on the outward journey. She resumes her journey at a slower speed. She spends time in Avonford and doesn't stop on the return journey.
(b) 1 hour (c) $1\frac{1}{4}$ km (d) 15 minutes (e) Check graph
(f) 10:45 am (g) $3\frac{1}{4}$ km

9 Negative numbers (pages 84–85)
1 Check answers.

2 (a) -1.5 (b) -7 (c) -2 (d) -3.5
(e) -4 (f) -6.35 (g) -0.5 (h) 0.6
3

A	B	$A + B$	$\dfrac{A + B}{2}$
-1	-2	-3	-1.5
-4	-10	-14	-7
-7	3	-4	-2
-12	5	-7	-3.5
-3.5	-4.5	-8	-4
-6.3	-6.4	-12.7	-6.35
-2.5	1.5	-1	-0.5
-1.7	2.9	1.2	0.6

4 (a) 4 (b) -34 (c) 1.6 (d) 2 (e) -5 (f) -90
(g) 6.1 (h) 9.7 (i) -24 (j) 60 (k) 1 (l) -1
(m) -6 (n) 3 (o) -3 (p) -1.2
5 (a) Total mark is ($5 \times$ number of correct answers)
$- (2 \times$ number of wrong answers)
(b)

	C	W	$5C - 2W$
Jane	10	10	30
Edward	12	8	44
Jessica	9	11	23
Davinda	5	15	-5

(c) (i) 93 (ii) 16
6 (a) 32 (b) -26 (c) -27 (d) -2.5

Activity
1 (a) $2, -2, 4, 2, -1$ (b) -1
2 2 left, 3 right, 6 left, 4 right, 4 right, 2 right
3 Ask your teacher to check your codes.

10 Fractions (pages 92–93)
1 $\frac{3}{4} = \frac{6}{8}$ $\frac{5}{6} = \frac{10}{12} = \frac{15}{18} = \frac{50}{60}$ $\frac{3}{7} = \frac{9}{21}$ $\frac{4}{5} = \frac{40}{50} = \frac{16}{20}$
$\frac{1}{3} = \frac{4}{12} = \frac{13}{39}$ $\frac{2}{4} = \frac{5}{10} = \frac{25}{50}$
2 (a) $\frac{2}{8}, \frac{3}{8}; \frac{3}{8}$ (b) $\frac{21}{35}, \frac{20}{35}; \frac{3}{5}$ (c) $\frac{9}{12}, \frac{10}{12}; \frac{5}{6}$ (d) $\frac{5}{9}, \frac{6}{9}; \frac{2}{3}$
(e) $\frac{6}{15}, \frac{4}{15}; \frac{2}{5}$ (f) $\frac{10}{15}, \frac{12}{15}; \frac{4}{5}$ (g) $\frac{15}{36}, \frac{14}{36}; \frac{5}{12}$ (h) $\frac{5}{30}, \frac{4}{30}; \frac{1}{6}$
3 (a) $\frac{55}{63}$ (b) $\frac{1}{6}$ (c) $\frac{3}{10}$ (d) $\frac{1}{2}$ (e) $\frac{5}{12}$ (f) $\frac{19}{24}$
(g) $\frac{8}{21}$ (h) $\frac{5}{24}$ (i) $\frac{1}{8}$ (j) $\frac{1}{16}$ (k) $\frac{1}{3}$ (l) $\frac{1}{5}$
4 (a) $1\frac{2}{5}$ (b) $1\frac{5}{24}$ (c) $1\frac{1}{15}$ (d) $1\frac{1}{3}$
(e) $1\frac{11}{24}$ (f) $1\frac{17}{24}$ (g) $1\frac{5}{9}$ (h) $1\frac{1}{12}$
5 (a) 50% (b) 75% (c) 10% (d) 30% (e) 5% (f) 35%
(g) 20% (h) 40% (i) 4% (j) 36% (k) 2% (l) 6%
6 Charles $\frac{4}{9}$ (44.4%) is bigger than $\frac{3}{7}$ (42.9%)
7

Activity	Number of children	As a fraction	As a percentage
Squash	9	$\frac{9}{75} = \frac{3}{25}$	12%
Computer games	18	$\frac{18}{75} = \frac{6}{25}$	24%
Tennis	15	$\frac{15}{75} = \frac{1}{5}$	20%
Orchestra	27	$\frac{27}{75} = \frac{9}{25}$	36%
Chess	6	$\frac{6}{75} = \frac{2}{25}$	8%
	Total 75	Total 1	Total 100%

8 (a) (i) $\frac{1}{4}$ (ii) $\frac{1}{3}$ (iii) $\frac{5}{24}$
(b) Red 30, Blue 10 (given), Green 40, Pink 25, White 5, Yellow 10
9 (a) $1\frac{1}{4}$ (b) $1\frac{11}{12}$ (c) $2\frac{16}{45}$ (d) $1\frac{11}{12}$
(e) $2\frac{43}{60}$ (f) $1\frac{5}{32}$ (g) $1\frac{1}{20}$ (h) $1\frac{3}{8}$
10 (a)–(e) Ask your teacher to check your answers.

11 Converting units (pages 100–101)
1 (a) 2 kg (b) 4.4 pounds
2 (a) (i) 18 kg (ii) 39.6 pounds (b) (i) 10 inches (ii) Less
3 (a) 3200 (b) 1800 litres
4 140 g flour, 30 g ground rice, 60 g castor sugar, 110 g butter
5 (a) Ask your teacher to check your graph
(b) (i) 160zl (ii) €5.00

6 (a) 0°C is not 0°F (b) 32°F (c) 99°F (d) 32°C (e) 68°F
7 (a) 18 litres (b) 6.7 kilograms (c) 48 kilometres
　(d) 9 ounces (e) 16°C (f) 17.8 pounds
8

256 km		
160 km	176 km	
64 km	240 km	112 km

9 (a) 1032 inches (b) 2580 cm (c) 25.80 m
10 (a) 636.36 kg (b) 0.64 tonnes (2 dp)

12 Shapes (pages 110–111)
1 (a) square (b) equilateral triangle (c) isosceles triangle
　(d) pentagon (e) scalene triangle
　(f) right-angled isosceles triangle (g) equilateral triangle
2 (a) Plan / Front elevation / Side elevation
　(b) Plan / Front elevation / Side elevation
　(c) Plan / Front elevation / Side elevation
3 Equilateral triangle, square, regular hexagon
4–6 Ask your teacher to check your diagrams.
7 (a) Square, rhombus (b) Square, rectangle
　(c) Square, rectangle, parallelogram, rhombus, kite, arrowhead
　(d) Square, rectangle, parallelogram, rhombus
　(e) Square, rectangle, parallelogram, rhombus
　(f) Square, rectangle (g) Square, rhombus, kite, arrowhead
　(h) Square, rectangle, parallelogram, rhombus, trapezium
　(i) Square, rhombus

Activity
Ask your teacher to check your design.

13 Prime factors, LCMs and HCFs (pages 118–119)
1 (a) 63 (b) 40 (c) 392 (d) 462 (e(539
2 (a) 2^3 (b) 3×5 (c) $2^2 \times 5$ (d) 2×5^2
　(e) $2 \times 5 \times 7$ (f) $2^4 \times 3 \times 5$ (g) $2^3 \times 3 \times 13$
　(h) 3^6 (i) 22×13 (j) $2^5 \times 7$ (k) $2^3 \times 5 \times 281$
3 (a) 6 (b) 10 (c) 4 (d) 30 (e) 60 (f) 60 (g) 360
　(h) 60 (i) 192 (j) 504 (k) 120 (l) 36
4 (a) 2 (b) 3 (c) 2 (d) 2 (e) 3 (f) 3 (g) 13 (h) 1
　(i) 14 (j) 11 (k) 18 (l) 9
5 (a) Brake fluid, change oil filter, tyres
　(b) Brake fluid, tyres, wiper blades, change timing belt
　(c) Every 108 000 km
6 (a) 6:03, 180 seconds later (b) 4 km (c) Midday
　(d) 20 minutes

Activity
1 2, making a total of 3. HCF of 6 and 9 is 3
2 5, making a total of 6. HCF of 12 and 18 is 6
3 360 ÷ old label   Check calculations

14 Doing a survey (pages 124–125)
Ask your teacher to check your survey.

15 Ratio and proportion (pages 132–133)
1 (a) 4:1 (b) 1:2 (c) 2:1 (d) 3:5 (e) 7:5
　(f) 2:3 (g) 10:9:6 (h) 3:2:4
2 (a) 2:1 (b) 50:3 (c) 1:3 (d) 2:7
　(e) 4:1 (f) 1:5 (g) 9:4 (h) 40:1
3 Tom 21, Kirsty 27
4 Mr Shaw €27 000, Mrs Taylor €12 000, Miss Weeks €9000
5 NSPCC €126, Save the Children Fund €210
6 24 m **7** (a) 100 m (b) 400 m **8** 54 km **9** €2600, €3400
10 (a) (i) $225:450$ (ii) $1:2$ (iii) $225:450:125$ (iv) $9:18:5$
　(b) (i) Not quite. They would be the same if the castor
　　　sugar were 112.5 g ($4\frac{4}{9}$ oz).
　　(ii) 1 ounce redcurrants, 2 ounces raspberries,
　　　$\frac{1}{2}$ ounce castor sugar (iii) 6 ounces redcurrants,
　　　12 ounces raspberries, 3 ounces castor sugar

11 (a) (i) $60:30:70:80$ (ii) $6:3:7:8$
　(b) 90°, 45°, 105°, 120°
　(c) Ask your teacher to check your pie chart.

16 Using formulae (pages 138–139)
1 (a) 8 (b) −290 (c) 10 (d) 7
2 (a) 130 (b) 7 (c) 700 (d) 42
3 (a) 10 (b) 5 (c) 0 (d) 0
4 No (112 km/h) **5** 9.4 ohms
6 (a) wl (b) $\frac{1}{2}wl$ (c) 3 cm
7 (a) 13 (b) 23 (c) $2n + 1$ (d) $4n - 1$
8 (a) $8 \times €9 + 5 = €77$ (b) $€(9n + 5)$ (c) €95
　(d) (i) 9 (ii) 7 (iii) 12

Activity
Ben, 74

17 Measuring (pages 146–147)
1 (a) 143 cm² (b) 23 cm² (c) 19.08 m² (d) $4\frac{1}{2}$ km²
　(e) 8.4 cm² (f) 25 cm² (g) 1.54 m² (h) 3605 mm² (i) $7\frac{9}{16}$ cm²
2 (a) m (b) m² or cm² (c) m³ (d) cm² or mm²
　(e) cm³ or mm³ (f) km
　(g) mm (h) m (i) m² (j) mm² (k) mm³ (l) cm²
3 (a) (i) 60 cm² (ii) 94 cm² (b) (i) 60 cm³ (ii) 94 cm²
　(c) (i) $275\frac{5}{8}$ m³ (ii) $283\frac{1}{2}$ m² (d) (i) 125 mm³ (ii) 150 mm²
4 Area of a rectangle = length × width　→ square units
　Perimeter of a rectangle = $2 \times (l + w)$　→ units
　Area of a triangle = $\frac{1}{2}$ base
　　× perpendicular height　→ square units
　Area of a parallelogram = base
　　× perpendicular height　→ square units
　Volume of a cuboid = length × width
　　× height　→ cubic units
　Surface area of a cuboid = $2lw + 2lh + 2wh$　→ square units
　Area of a trapezium = $\frac{1}{2} \times (a + b) \times h$　→ square units
5 (a) 11.6 m (b) 19.6 m (c) 16.065 m³
6 (a) 7100 cm² (b) 51 (c) No
7 (a) 8.9 cm × 9.3 cm × 6 cm (b) 383.94 cm² (c) 496.62 cm³

Investigation
1 8
2 1×144 has greatest perimeter.
　12×12 has the least perimeter.
3 All have the same area.

18 Percentages (pages 154–155)
1 (a) 65% (b) 68% (c) 70% (d) $92\frac{1}{2}$% (e) 60%
2 (a) $\frac{1}{5}$ (b) $\frac{3}{10}$ (c) $\frac{4}{5}$ (d) $\frac{1}{8}$
3 (a) 40% (b) 22% (c) 6% (d) $2\frac{1}{2}$%
4 55% **5** 7 **6** 3 **7** 9.2 g **8** €157.50
9 (a) €45 (b) €132 (c) €177 (d) €27
10 30% **11** 20% **12** 2%
13 (a) €18 (b) €38.40 (c) €336 (d) €15
14 (a) €6 (b) €247.50 (c) €3360 (d) €2268
15 (a) €270 (b) €212.50 (c) €7.50
16 (a) (i) 8% (ii) 8% (b) (i) 93% (ii) 93%

19 Symmetry (pages 162–163)
1 (a) (i) ... (ii) ... (iii) ... (iv) ...
　(v) ... (vi) ... (vii) ... (viii) ...
　(b) (i) 2 (ii) 3 (iii) 2 (iv) 2 (v) 1 (vi) 2 (vii) 1 (viii) 1
2 (a)(i) (b)(i) (c)(i) (d)(i) (e)(i)
　(a)(ii) 1 (b)(ii) 2 (c)(ii) 2 (d)(ii) 1 (e)(ii) 6
3 (a) (b)

Activity
Ask your teacher to check your snowflakes.

Investigation
Ask your teacher to check your masks.

20 Equations (pages 172–173)
1 (a) $p = 4$ (b) $t = 14$ (c) $a = 2$ (d) $r = 5$
   (e) $y = 6$ (f) $m = 4.5$ (g) $k = 18$ (h) $b = 63$
2 (a) $n = 2$ (b) $d = 5$ (c) $f = 5$ (d) $h = 1$
   (e) $t = 3$ (f) $p = 2$ (g) $k = 4$ (h) $r = 13$
3 (a) $t = 10$ (b) $n = 2$ (c) $g = 5$ (d) $p = 3$
   (e) $x = 7$ (f) $r = 1$ (g) $h = 6$ (h) $k = 1$
4 (a) $3f + 3 = 12$ (b) €3 (c) $3 \times €3 = €9$, $€9 + €3 = €12$
5 (a) (i) $6p + 16$ (ii) $4p + 44$ (b) $6p + 16 = 4p + 44$
   (c) 14 cent (d) €1
6 (a) $4m + 5 = 2m + 6$ (b) 0.5 kg
   (c) $4 \times 0.5 + 5 = 7$ and $2 \times 0.5 + 6 = 7$ (d) 7 kg
7 (a) Susan (b) Check working

Activity
Ask your teacher to check your cards.

21 Probability (pages 180–181)
1 (a) $\frac{1}{8}$ (b) $\frac{7}{8}$ (c) $\frac{3}{8}$ (d) $\frac{1}{2}$ (e) $\frac{1}{4}$ (f) $\frac{3}{4}$
2 (a) $\frac{17}{64}$ (b) $\frac{47}{64}$ (c) $\frac{5}{64}$ (d) $\frac{1}{8}$ (e) $\frac{4}{64}$ or $\frac{1}{16}$ (f) $\frac{17}{64}$
3 (a) 95 (b) (i) $\frac{43}{95}$ (ii) $\frac{18}{95}$
   (c) (i) fairly reliable
       (ii) not very reliable (different time of day)
4 Check estimates, explanations and scale.
5 (a)

×	0	1	2	3
0	0	0	0	0
1	0	1	2	3
2	0	2	4	6
3	0	3	6	9

   (b) $\frac{7}{16}$ (c) $\frac{1}{8}$ (d) $\frac{3}{16}$
6 (a) K♠ Q♠   Q♠ J♠   J♠ K♥   K♥ Q♥   Q♥ J♥
       K♠ J♠   Q♠ K♥   J♠ Q♥   K♥ J♥
       K♠ K♥   Q♠ Q♥   J♠ J♥
       K♠ Q♥   Q♠ J♥
       K♠ J♥
   (b) $\frac{1}{5}$ (c) $\frac{3}{5}$ (d) $\frac{1}{15}$ (e) $\frac{4}{15}$

22 Number patterns (pages 190–191)
1 (a)

Pattern 4   *Pattern 5*

   (b)

Pattern	1	2	3	4	5
Number of matches	4	7	10	13	16

   (c) 3 (d) (i) 19 (ii) 31 (iii) 301   (e) $M = 3P + 1$
   (f) (a)

Pattern 4   *Pattern 5*

   (f) (b)

Pattern	1	2	3	4	5
Number of matches	3	5	7	9	11

   (f) (c) 2
   (f) (d) (i) 13 (ii) 21 (iii) 201   (f) (e) $M = 2P + 1$
2 (a) (i) 4 (ii) 3
   (b) (i) (i) 17 (i) (ii) 37 (i) (iii) 77
       (ii)(i) 22 (ii)(ii) 37 (ii)(iii) 67
   (c) (i) $M = 4P - 3$ (ii) $M = 3P + 7$
3 (a) 7, 10, 13, 16, 19 (b) 18, 16, 14, 12, 10 (c) 5, 8, 13, 20, 29
4 (a) 18, 22, 26 (b) 78 (c) $4n - 2$ (d) 198

Activity
Ask your teacher to check your answers.

23 Multiplying and dividing fractions (pages 196–197)
1 *See table at the top of next column.*
2 (a) 15 (b) 8 (c) 15 (d) 4 (e) 16 (f) €0.75
   (g) €2.66 (h) €2.88
3 (a) $\frac{1}{6}$ m² (b) $\frac{5}{35}$ m² (c) $\frac{15}{128}$ m²
4 (a) $\frac{1}{10}$ m² (b) $\frac{5}{24}$ m² (c) $\frac{7}{64}$ m²

×	$\frac{1}{2}$	$\frac{3}{7}$	$\frac{7}{9}$	$\frac{3}{5}$	$\frac{1}{5}$	$\frac{3}{4}$	$\frac{1}{3}$
$\frac{1}{2}$	$\frac{1}{4}$	$\frac{3}{14}$	$\frac{7}{18}$	$\frac{3}{10}$	$\frac{1}{10}$	$\frac{3}{8}$	$\frac{1}{6}$
$\frac{3}{4}$	$\frac{3}{8}$	$\frac{9}{28}$	$\frac{7}{12}$	$\frac{9}{20}$	$\frac{3}{20}$	$\frac{9}{16}$	$\frac{1}{4}$
$\frac{2}{5}$	$\frac{1}{5}$	$\frac{6}{35}$	$\frac{14}{45}$	$\frac{6}{25}$	$\frac{2}{25}$	$\frac{3}{10}$	$\frac{2}{15}$
$\frac{5}{8}$	$\frac{5}{16}$	$\frac{15}{56}$	$\frac{35}{72}$	$\frac{3}{8}$	$\frac{1}{8}$	$\frac{15}{32}$	$\frac{5}{24}$
$\frac{6}{7}$	$\frac{3}{7}$	$\frac{18}{49}$	$\frac{2}{3}$	$\frac{18}{35}$	$\frac{6}{35}$	$\frac{9}{14}$	$\frac{2}{7}$
$\frac{5}{16}$	$\frac{5}{32}$	$\frac{15}{112}$	$\frac{35}{144}$	$\frac{3}{16}$	$\frac{1}{16}$	$\frac{15}{64}$	$\frac{5}{48}$
$\frac{1}{3}$	$\frac{1}{6}$	$\frac{1}{7}$	$\frac{7}{27}$	$\frac{1}{5}$	$\frac{1}{15}$	$\frac{1}{4}$	$\frac{1}{9}$

5 (a) $\frac{2}{3}$ m (b) $\frac{2}{5}$ m (c) $1\frac{1}{4}$ m (d) 1 m (e) $1\frac{1}{2}$ m (f) $\frac{1}{2}$ m
6 (a) (i) $\frac{1}{6}$ (ii) $\frac{1}{8}$ (iii) $\frac{1}{3}$
   (b) Entertainment €100, sweets and snacks €75,
       clothes €200, savings €150, CDs €75
7 (a) $1\frac{1}{15}$ km h^{-1} (b) $\frac{3}{8}$ km h^{-1} (c) $1\frac{7}{8}$ km h^{-1} (d) 60 km h^{-1}

24 Transformations (pages 204–205)
1 Ask your teacher to check your answers.
2 (a) F to G or G to F (b) J to K or K to J (c) B to F or F to B
3 (a) (0, 0) (0, 5) (3, 5) (3, 4) (1, 4) (1, 3)
       (2, 3) (2, 2) (1, 2) (1, 1) (3, 1) (3, 0)
   (b) $x = 7\frac{1}{2}$, $x = 11\frac{1}{2}$, $x = 15\frac{1}{2}$, $x = 19\frac{1}{2}$, $x = 23\frac{1}{2}$
   (c)–(g) Ask your teacher to check your diagram.
   (h) Pattern: E, reflection of E, E, reflection of E, E
4 22 to the left and 56 up
5 (a)–(c) Check diagrams
   (d) Enlargement, scale factor $1\frac{1}{2}$, centre C
6 45°, Yes, also 90°, 135°, 180°, …

Investigation
1 Rotation 360° clockwise or anticlockwise about centre
2 Scale factor is 1
3 Yes
4 0 to the right and 0 down

25 Getting the most from your calculator (pages 212–213)
1 (a) 64 (b) 22.22… (c) 57 289 761 (d) 37.647 68 (e) 1.2 (f) 0.3
2 (a) (i) $42 \times (87 - 63)$ (ii) 1008
   (b) (i) $14 \times (65 + 18)$ (ii) 1162
   (c) (i) $(8.7 + 3.9)^4$ (ii) 25 204.7376   (d) (i) $\frac{3}{4} + 2\frac{2}{3}$ (ii) $3\frac{5}{12}$
   (e) (i) $(2.9 - 1.5) \div (1.9 + 0.6)$ (ii) 0.56
   (f) (i) $2.7 - \dfrac{1.3}{(1.9 + 0.6)}$ (ii) 2.18
   (g) (i) $5^7 \div 5^6$ (ii) 5   (h) (i) $(\frac{2}{3})^4$ (ii) 0.197 530 864
3 For (c) (i) [87] [x^y] [4] [=]   (ii) Check calculation
   For (f) (i) [0.027] [SHIFT] [y^x] [3] [=]   (ii) Check calculation
       or [3] [SHIFT] [y^x] [0.027] [=]
4 (a) (i) 0.75 (ii) 0.333… (iii) 0.375 (iv) 0.07 (v) 0.0125
   (b) (i) $\frac{1}{10}$ (ii) $\frac{1}{8}$ (iii) $\frac{1}{25}$ (iv) $\frac{3}{80}$ (v) $\frac{1}{200}$
5 (a) Subtract 5, take square root
   (b) (i) add 6, take cube root, add 8
       (ii) multiply by 5, then cube it
6 (a)

Number	Square	Last digit
1	1	1
2	4	4
3	9	9
4	16	6
5	25	5
6	36	6
7	49	9
8	64	4
9	81	1

   (b) and (c) 0, 1, 4, 9, 6, 5, 6, 9, 4, 1 and continue to repeat
       this pattern.
   (d) 2, 3, 7, 8 (e) Because it ends with a 7

Investigation
1 (a) -1 (b) 1 (c) -1
2 (a) -1 (b) $(-1)^{\text{odd}} = -1$ $(-1)^{\text{even}} = +1$